Routledge Revivals

Petroleum Company Operations and Agreements in the Developing Countries

Originally published in 1984, this study focuses on petroleum agreements between non-OPEC LDCs with oil-importing LDCs and how issues such as high oil prices affect each country. The information presented in this study was drawn from interviews with petroleum officials in petroleum companies, petroleum ministries and unpublished documents such as contracts and focussing on case studies of countries such as Peru, Guatemala and Malaysia. This title will be of interest to students of environmental studies and economics.

Petroleum Company Operations and Agreements in the Developing Countries

Raymond F. Mikesell

RFF PRESS
RESOURCES FOR THE FUTURE

First published in 1984
by Resources for the Future, Inc.

This edition first published in 2015 by Routledge
2 Park Square, Milton Park, Abingdon, Oxon, OX14 4RN
and by Routledge
711 Third Avenue, New York, NY 10017

Routledge is an imprint of the Taylor & Francis Group, an informa business

© 1984 Resources for the Future, Inc.

Publisher's Note
The publisher has gone to great lengths to ensure the quality of this reprint but
points out that some imperfections in the original copies may be apparent.

Disclaimer
The publisher has made every effort to trace copyright holders and welcomes
correspondence from those they have been unable to contact.

A Library of Congress record exists under LC control number: 83043265

ISBN 13: 978-1-138-18499-2 (hbk)
ISBN 13: 978-1-315-64477-6 (ebk)

PETROLEUM COMPANY OPERATIONS AND AGREEMENTS IN THE DEVELOPING COUNTRIES

Raymond F. Mikesell

RESOURCES FOR THE FUTURE / WASHINGTON, D.C.

Published by Resources for the Future, Inc., 1755 Massachusetts Avenue, N.W.,
Washington, D.C. 20036
Resources for the Future books are distributed worldwide by The Johns
Hopkins University Press.

Library of Congress Cataloging in Publication Data

Mikesell, Raymond Frech.
   Petroleum company operations and agreements in the developing countries.

   Bibliography: p.
   Includes index.
   1. Petroleum industry and trade—Government policy—Developing countries. 2. Petroleum
industry and trade—Law and legislation—Developing countries. 3. Corporations, Foreign—
Government policy—Developing countries. 4. Revenue—Developing countries. I. Title.
HD9578.D44M55 1984   338.2'7282'091724   83-43265
ISBN 0-915707-07-1

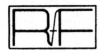

Contents

II   CASE STUDIES OF MODERN CONTRACTUAL ARRANGEMENTS   57

5   Indonesia's Petroleum Agreements: The Evolution of the Production-Sharing Contract (PSC)   59

6   Peru's Production-Sharing Contracts   69

7   Guatemala's Production-Sharing Contracts   77

8   The Production-Sharing Contracts of Malaysia, the Philippines, and Egypt   86

Preface

This book reflects my continued interest and research on petroleum agreements which began in the 1940s when I published (with Hollis B. Chenery) *Arabian Oil* (University of North Carolina Press, 1949). The emphasis in this study on petroleum contracts negotiated by non-OPEC LDCs reflects the widespread concern for the hardship imposed on the oil-importing LDCs by the several-fold rise in oil prices, and the international efforts to reduce the dependence of these countries on imported petroleum. A major issue is whether the contracts offered by the governments of these countries to international petroleum companies have been sufficiently attractive to induce the amount of exploration and development warranted by their petroleum potential as determined by geological conditions. It is hoped that this study will be of value to officials of both host governments and petroleum companies, and to students and practitioners of development and resource economics generally.

Most of the information for the country case studies was derived from interviews with officials of petroleum companies and petroleum ministries in the countries studied, and from a large number of unpublished documents, some of which are confidential. This study would not have been possible without the possession of a substantial library of petroleum agreements made available from a variety of sources, including international petroleum companies, petroleum ministries, the UN Centre for Transnational Corporations, the Institute for Foreign and International Law at Frankfurt am Main, and The Barrows Company, Inc. of New York.

In addition to financial assistance from Resources for the Future, Inc., this study has benefited from my work under two contracts from the U.S. Department of Energy for the preparation of reports on petroleum agreements in several Latin American countries. I also visited several Asia-Pacific countries under a grant from the Fund for Multinational Management Education to study both mining and petroleum agreements. In the case of all my visits to foreign countries, I received generous support from the economic counselors of the American embassies who arranged contacts for me with petroleum ministries and provided me with background information. Among the petroleum companies whose officials provided information both in the United States and abroad were Amoco, Arco, Belco Petroleum, Caltex, Cities Service, Exxon, Getty Oil, Gulf Oil, Mobil Oil, Petromaya (Guatemala), Occidental, Royal Dutch Shell, Superior Oil, Standard Oil of California, Stanvac of Indonesia, Texaco, and Union Oil.

Among individuals to whom I owe a special debt of gratitude are Gordon Barrows of The Barrows Company, Inc., New York; Raymond T. Adams, Superior Oil; Dennis O'Brien of the Department of Energy; Hernan Mejia of the Central Bank of Colombia; William S. Pintz, East-West Center, Honolulu; Napier Collyns, Scallop Corporation; Keith F. Palmer and F. Z. Jasperson, World Bank; Erich Schanze, University of Frankfurt; John Whitney, Whitney and Whitney, Reno; Walter Levy, W. J. Levy Consultants, New York; John E. Kirby, Esso-Eastern, Houston; and my colleague at the University of Oregon, M. A. Grove. Finally, this book would not have been possible without the editing and reference services of my wife, Irene, and the secretarial skills of Letty Fotta.

February 1984   R.F.M.

1

Introduction

This study of international petroleum operations and agreements focuses mainly on the non-OPEC LDCs for the following reasons. First, most of the non-OPEC LDCs are net oil importers. The high cost of oil imports is retarding their economic development and has been a substantial factor in the rise in their external indebtedness. Second, the non-OPEC LDCs are in general in greater need of attracting international petroleum companies for exploration and development than are the large OPEC producers, most of which have very large proven reserves, have nationalized their petroleum industries, and have developed a substantial degree of domestic technical capacity for their petroleum operations. Third, although it seems unlikely that many (if, indeed, any) of the oil-importing, developing countries (OIDCs) will become major petroleum exporters to the United States and other western countries, the achievement of a higher degree of self-sufficiency in the OIDCs as a group would reduce dependence on a handful of large OPEC producers, thereby diminishing their monopoly power. Finally, the discovery and development of oil reserves in developing countries outside the Middle East

that are actual or potential oil exporters will also reduce the dependence of western countries on Middle East oil.

The division between OPEC and non-OPEC LDCs is a somewhat artificial one from the standpoint of the major interests of this study. For example, Mexico, a non-OPEC country, is a larger producer and exporter of petroleum than several of the OPEC countries, but Mexico nationalized its petroleum industry in 1938 and has no interest in negotiating contracts for equity investment by private petroleum companies. On the other hand, Indonesia (an OPEC member), unlike the large Middle East producers such as Saudi Arabia, Iraq, Iran, and Kuwait, has not nationalized its petroleum industry, is quite anxious to negotiate new contracts with international petroleum companies in order to expand its reserves and production, and constitutes a relatively secure source of oil outside the Middle East. For these reasons, and because Indonesia has been a pioneer in the development of new types of petroleum agreements, Indonesian contracts are analyzed in one of the case studies covered in this survey.

There are wide differences among the non-

1

OPEC LDCs with respect to their current petroleum export–import balances and their likely production potential. A few countries, including Egypt and Malaysia, are substantial petroleum exporters and are, therefore, dependent upon the development of their petroleum resources for increasing their foreign exchange income. Another group of countries is in near balance in petroleum production and consumption, but must expand its output substantially over the next few years if the countries are to avoid becoming large net importers. This group includes Argentina (net importer), Peru (net exporter), and Zaire (net exporter). Still another group, including Brazil, Colombia, and India, has substantial production and good prospects, but is a considerable distance from self-sufficiency. The largest group of LDCs has no production and for most of them the outlook for significant petroleum discoveries is not promising.

Regardless of the petroleum exploration efforts in the OIDCs, there is virtually no chance of their becoming (as a group) self-sufficient in petroleum in the near future or perhaps ever on the basis of present geological knowledge. The World Bank staff, which is perhaps more optimistic than other investigators with respect to the petroleum potential of the OIDCs, projects a rise in production in these countries from 2.0 million barrels per day (bpd) in 1980 to a range of 3.6 to 4.8 million bpd in 1990, depending upon the intensity of petroleum effort. Taking into account the World Bank's petroleum demand projections in these countries, the OIDCs would increase their net imports from 4.5 million bpd in 1980 to a range of 6.4 to 7.6 million bpd in 1990, depending upon the success of their petroleum programs.[1] In 1980 the OIDCs paid $49 billion for their oil imports. Assuming a price of $30 per barrel and an increase of 3 percent per year in real terms, the World Bank estimates that the oil bill for the OIDCs in 1990 would range from $93 billion to $111 billion for the two oil import projections. However, this

assumes that these countries would have the foreign exchange to finance such a level of oil imports. It seems more likely that the OIDCs will be forced to curtail their energy requirements and to shift to alternative energy sources, even in the case of the maximum petroleum production projections.

Exploration and development activity is a function of many variables, including the basic geologic potential and the accessibility of the exploration areas. Aside from the physical factors, an important variable is the commitment of governments to petroleum investment, either directly through a government oil enterprise (GOE) or by inducing foreign petroleum companies to invest. Before the sharp rise in world petroleum prices in 1973–74, many OIDCs had little interest in promoting oil exploration by negotiating contracts with petroleum companies or by providing adequate financial and technical resources for their GOEs. Following the rise in oil prices, some of the OIDCs, such as Brazil and India, expanded exploration and development activities of their GOEs, but did not encourage foreign petroleum investment until recently.

The nature of petroleum contracts offered to foreign investors is only one of the factors, other than geologic potential, that determines the level of foreign exploration activities. Political and economic stability and the history of the country's relations with foreign investors also play a role in attracting petroleum companies.[2] However, the principal focus of this study is an analysis of the implications of petroleum contract terms and host country fiscal systems for attracting foreign petroleum companies and for maximizing production and revenue of the host countries. It seeks to throw light on several important issues, although an adequate treatment of some of them is beyond the terms of this study. The major issues are as follows:

---

[1]World Bank, *Energy in Developing Countries* (Washington, D.C., August 1980), pp. 15 and 20. A more recent World Bank report, *The Energy Transition in Developing Countries* (April 1983), projects net imports of oil by the OIDCs of only 5.9 million bpd in 1995 (p. 11).

[2]The determinants of foreign petroleum investment in non-OPEC developing countries, including contractual arrangements, fiscal regimes, political risk, and the behavioral patterns of petroleum companies, are the subject of an empirical study currently being undertaken by Harry G. Broadman and Joy Dunkerley at Resources for the Future, Inc., Washington, D.C.

1. *What are the effects of various types of fiscal arrangements on rates of return and risk for petroleum investments and upon the efficiency of petroleum company operations from the standpoint of maximizing oil discoveries, production, and host-government revenues?*

This question requires an economic analysis of the impacts that various types of taxes and production and revenue-sharing arrangements found in modern petroleum agreements have on petroleum company rates of return and on investment and production decisions. The conclusions from this analysis, presented in chapter 4, provide the criteria for evaluating the petroleum agreements employed by individual countries presented in part II.

2. *How should the petroleum contracts offered by the governments of non-OPEC LDCs be evaluated with respect to their attractiveness for international petroleum companies, and their effects on the efficiency of petroleum company operations in maximizing oil discoveries and production?*

Each of the petroleum agreements examined in the case studies embodies a unique mix of fiscal arrangements that are of varying attractiveness to petroleum companies and that have different impacts on the efficiency of their operations. The relative attractiveness of petroleum contracts cannot be adequately measured because of differences in geologic conditions and in the political and economic environment among countries, as well as differences in the behavioral patterns of petroleum companies in response to conditions in the host countries. Also, once several significant oil discoveries are made in a country, there is something of a herd effect on the part of international oil companies, whose expectations are changed by the discoveries. Nevertheless, there is evidence that, under a given set of physical and environmental conditions, changes in contract terms affect the interest of international petroleum companies in negotiating contracts. The analysis of the effects of specific contract terms on petroleum company operations is much more difficult because we do not know the bases for decisions affecting the exploration and development of marginal

fields. Simulating the effects of specific fiscal regimes on expected rates of return on investments in areas with differing geologic prospects makes it possible to infer the responses of profit-maximizing firms.

3. *Do existing petroleum agreements give too large a share of the economic rent to the petroleum companies or to the host governments?*

The proper division of the economic rent, from the standpoint of maximizing host country revenues, depends on geologic conditions in the host country and whether important discoveries have been made. In addition, there are a number of tradeoffs in the formulation of contract terms and fiscal arrangements that a host country should keep in mind. These include the tradeoff between risk sharing and net profit sharing; between maximization of production and the share of revenues from marginal fields; between smaller host government revenue in the near term and larger host government revenue in the longer term; and between a larger share of net revenues from a few contracts covering a small portion of the areas with petroleum potential versus a smaller share of the net revenues from a larger number of contracts covering a large portion of the areas with petroleum potential.

4. *What should be the role of international assistance agencies in transferring financial and technical resources to the governments of non-OPEC LDCs?*

The World Bank and the United Nations have been providing substantial financial and technical assistance to GOEs in recent years. Some of these programs are highly controversial and have been criticized on the grounds that they subsidize activities that could be undertaken more efficiently by international petroleum companies. Is this criticism justified?

5. *In view of the interest of industrial countries, such as the United States, in promoting development of petroleum potential in non-OPEC LDCs, how might parent-country tax and other policies be adjusted to promote investments by international petroleum companies in foreign operations?*

The issue of U.S. government tax policies and foreign investments of U.S. petroleum companies is highly complex in terms of conflicting public interests and the impact of tax arrangements on the overseas operation of U.S. petroleum companies. A related issue is the role of investment insurance programs for petroleum investments in the LDCs.

Part I of this book consists of three chapters that provide background and analysis relevant for the case studies in part II. Chapter 2 examines the potential for petroleum production in non-OPEC LDCs, both overall and by major countries. Chapter 3 briefly reviews the evolution of contracts between petroleum companies and LDCs and describes the major forms of contracts. Chapter 4 presents a theoretical analysis of the effects of various types of taxes and other contract provisions on investment decisions by petroleum companies, on the efficiency of their operations, and on the financial returns to the host governments. The analysis is supplemented by simulations of the effects of different fiscal regimes on hypothetical petroleum fields.

The seven chapters in part II analyze petroleum agreements in nine countries and evaluate them in terms of the findings and criteria formulated in chapter 4. Together the case studies cover the principal types of petroleum agreements in operation in the LDCs and illustrate the issues that arise in the negotiation and implementation of contracts between host governments and international petroleum companies.

Most of the case studies reflect interviews by the author with government officials, petroleum company representatives, and independent observers in the host countries. The analysis is based on a large number of actual petroleum agreements collected from a variety of sources, and, in some cases, from model agreements prepared by the governments of individual countries. Many of these agreements have never been published or officially made public, and some are confidential.[3]

The chapters in part III review policies and activities of international agencies and the U.S. government with respect to petroleum operations in developing countries, including their influence on foreign investment. Chapter 11 reviews the tax policies of the U.S. government with respect to U.S. investment in petroleum in the LDCs, while chapter 12 examines the activities of United Nations agencies and the World Bank Group in assisting non-OPEC LDCs with exploration and development of their resources. Chapter 13 summarizes the conclusions on petroleum agreements and on the roles of the U.S. government and international agencies in promoting petroleum production in the non-OPEC LDCs, and provides some policy recommendations.

[3]Copies of a number of petroleum agreemsnts are available from The Barrows Company, P.O. Box 1591, Grand Central Station, New York, N.Y. 10022. The United Nations Centre for Transnational Corporations also has a fairly complete set of petroleum agreements.

I

GENERAL ANALYSIS

2

The Petroleum Potential of Non-OPEC Countries

Reserves and Ultimately Recoverable Oil Resources

Petroleum geologists do not agree about the world's petroleum potential or that of the non-OPEC countries with which this book is mainly concerned. There is agreement on the volume of *proved* reserves—those oil deposits whose existence is known with a high degree of confidence on the basis of more or less standardized geologic and engineering techniques and which are commercially producible.[1] A more fundamental question for this study is: How much ultimately recoverable oil resources likely to be commercial in the foreseeable future are there in the non-OPEC countries? This chapter also considers the prospects for successful petroleum exploration in individual countries and whether sufficient exploration is being undertaken in the light of these prospects.

The bulk of the world's recoverable resources are classed as undiscovered potential resources; their existence is based on geologic studies of the world's sedimentary basins, of which some 600 have been identified. Some of these basins are known to be productive or capable of production, while others have been explored to some degree without the discovery of reserves. Still others are essentially unexplored, either because of their location or because they are not believed to be promising.

As of 1982, the proved reserves of the non-OPEC LDCs were estimated at 79 billion barrels, or about 14 percent of the proved reserves of the non-Communist world. Over 70 percent of these reserves are in Latin America (about 56 billion barrels) and 48 billion barrels were accounted for by Mexico.[2] Non-OPEC LDC

---

[1] In addition to *proved* reserves, there are *probable* reserves whose existence is less well known and which are expected to be commercial in the future. There is also a portion of *discovered* (but not proved) resources that is noncommercial, but which conceivably might become commercial if petroleum prices were to rise sufficiently in relation to the costs of extraction. These are sometimes called *static* resources. *Discovered* resources include cumulative production to date, commercial reserves (which are divided into *proved* reserves and *probable* reserves), and static resources. *Discovered* resources are based on an analysis of data derived from existing oil fields, including extrapolations, theoretical calculations, and comparisons with other similar fields for which more data are available. See *How Much Oil and Gas?* Exxon Background Series (New York, May 1982), p. 5.

[2] *Oil and Gas Journal* (December 28, 1981) pp. 86–87.

7

reserves for Africa are estimated at 9.5 billion barrels; for the Middle East, 4.8 billion; and for Asia-Pacific, 8.8 billion (table 2-1). About 14 billion barrels, or 18 percent of the non-OPEC LDC reserves, are in the oil-importing, developing countries. Existing reserves will largely determine output for the next decade because of the time required to explore and develop newly discovered resources.

Table 2-1. Estimated Proved Reserves and Crude Oil Production in Non-OPEC Developing Countries, 1982

	Production (1,000 bpd)	Reserves (billions of barrels)
Africa		
Angola	122	1.63
Cameroon	109	0.53
Congo	87	1.55
Egypt	667	3.32
Ivory Coast	9	0.11
Tunisia	106	1.86
Zaire	22	0.14
Other	2	0.40
Total	1,124	9.54
Asia-Pacific		
Brunei	155	1.24
Burma	32	0.03
India	384	3.42
Malaysia	306	3.32
Pakistan	12	0.02
Philippines	7	0.36
Other	8	0.22
Total	904	8.61
Middle East		
Bahrain	45	0.20
Oman	328	2.73
Syria	175	1.52
Turkey	45	0.28
Other	1	0.10
Total	594	4.83
Western Hemisphere		
Argentina	483	2.59
Bolivia	24	0.18
Brazil	252	1.75
Chile	41	0.76
Colombia	140	0.54
Guatemala	6	0.05
Mexico	2,734	48.30
Peru	198	0.84
Trinidad and Tobago	182	0.58
Other	1	0.01
Total	4,061	55.60

Source: *Oil and Gas Journal*, December 27, 1982, pp. 78–79. Excludes China and the Soviet bloc countries.

Estimates of ultimately recoverable world oil resources are quite speculative and vary from the 2,600 to 6,500 billion barrel range estimated by Bernardo Grossling[3] to that provided by Richard Nehring of 1,700 to 2,300 billion barrels.[4] Nehring's maximum estimate is in general agreement with the mean of the estimates of total oil and gas resources by a group of 27 petroleum geologists given below, after adjustment for the natural gas portion. Although he does not provide an estimate of ultimately recoverable crude resources for non-OPEC LDCs, his regional breakdown suggests that the worldwide distribution of ultimately recoverable oil resources does not differ substantially from the distribution of presently known reserves.

At the 1977 World Energy Conference, 27 petroleum geologists representing petroleum companies, government agencies, and independent consultants submitted estimates of the world oil and gas resource base, including what had been produced, the rest of what had already been discovered, and what is still undiscovered. The estimates range from 3,000 to 7,000 billion barrels of oil equivalent. Statistical analysis of the experts' projections indicates that the mean or most likely result is about 4,500 billion barrels (oil equivalent) with a 95 percent probability that the resource base is at least 3,000 billion barrels and only a 5 percent probability that it could be as high as 6,500 billion barrels. From this amount, cumulative production to date of 700 billion barrels would need to be subtracted to arrive at total resources discovered plus undiscovered potential. On the basis of industry projections alone, Exxon geologists arrived at a total oil and gas resource base of 3,000 to 5,000 billion barrels, including both total discovered resources and undiscovered potential.[5] However, on the basis of the ratio of proved

[3]U.S. Geological Survey, *Survey Circular 724* (Washington, D.C., USGS, 1974).

[4]Richard Nehring, *Giant Oil Fields and World Oil Resources* (Santa Monica, Calif., Rand Corp., June 1978), p. 88. Nehring reduced this estimate to 1,600 to 2,000 billion at a UNITAR conference in Montreal, December 1979, as reported by Office of Technology Assessment, *World Petroleum Availability, 1980–2000* (Washington, D.C., Government Printing Office, October 1980).

[5]Exxon, *How Much Oil?*, pp. 9–10.

natural gas reserves to total proved oil and gas reserves in 1982, about 40 percent of this resource base would represent gas; therefore, the range for petroleum would be 1,800 to 3,000 billion barrels. If the reserve base of the non-OPEC countries is assumed to be proportional to these countries' share of the world's proved reserves, the oil reserve base of the non-OPEC countries would be 250 to 420 billion barrels, over 60 percent of which would be accounted for by Mexico. Thus, estimated non-OPEC oil and gas resources, excluding Mexico, are only 100–170 billion barrels of oil.

Special interest is attached to the reserve potential of the OIDCs and to the possibility of discovering oil in nonproducing countries or of expanding production through increased exploration and development in those that are producing. A U.S. Department of the Treasury study gives a range of 55 to 100 billion barrels of oil as the ultimately recoverable resources of these countries, of which 21 to 47 billion barrels are in South America and 20.5 to 38.5 billion barrels are in Africa. The largest oil-producing potential in the OIDCs is in the following countries: Argentina, Brazil, Colombia, India, Burma, Ghana, Somalia, Sudan, Guatemala, and Ivory Coast.[6] This resource potential is substantial, although it tends to be concentrated in those OIDCs where oil has been discovered and where nearly 2 million bpd was produced in 1982. In addition, most of the exploration in recent years has taken place in the oil-producing OIDCs as contrasted with that in the non-oil producers.

Since there are a number of LDC importing countries that have not been adequately explored for one reason or another, the view is sometimes expressed that huge unknown oil resources may exist in some of these countries. However, this viewpoint is not held by most petroleum geologists, who are quite pessimistic about the possibility of discovering many, or even any, supergiant fields (with over 5 billion barrels) such as those found in the Middle East and in Venezuela, Mexico, Texas, and the North Slope of Alaska. Only one supergiant field has been found since the 1960s (Mexico), and the discoveries of larger fields have been decreasing while the number of discoveries and volume of reserves in smaller fields (under 50 million barrels) have been increasing.[7]

Exploration

A substantial amount of exploration activity, including seismic surveys, exploratory drilling, and other types of exploration, has taken place in recent years in the non-OPEC LDCs. Of the 113 non-OPEC developing countries and semi-independent territories examined in a 1978 Exxon survey, exploratory drilling took place in 71 of these countries during 1967–76; seismic surveys and other exploration were carried out in 22; while only 20 countries were not explored at all. These 20 are mostly small island nations or countries where geologic prospects are poor. About 13,000 exploratory wells were drilled in the world outside the United States, Canada, and the Communist countries during 1967–76, of which 24 percent were drilled in the industrial countries, 26 percent in the OPEC countries, and the other half in non-OPEC LDCs. Of the 6,500 exploratory wells drilled in the non-OPEC LDCs over the 1967–76 period, the vast majority (5,416) were in the 16 countries that had significant discoveries prior to 1967. Another 851 exploratory wells were drilled in 25 countries regarded as "encouraging," and only 234 exploratory holes were drilled in 30 countries where no positive encouragement was indicated.[8]

A 1981 World Bank report points out that during the 1970s drilling effort increased considerably more in the industrial countries than in the non-OPEC LDCs. Between 1970–72 and 1976–78, drilling effort (all drilling measured in feet per year) for the industrial countries increased by nearly 66 percent. For the non-OPEC

---

[6]Office of International Energy Policy, *The World Bank Energy Lending Program* (mimeo) (Washington, D.C., U.S. Department of the Treasury, July 28, 1981), pp. 3–5. Resource estimates in this report were provided by the U.S. Department of Energy.

[7]Exxon, "Exploration in Developing Countries," (unpublished study) (New York, June 1978).

[8]Ibid., chart 9.

LDC oil exporters, the corresponding figure was only 1.0 percent, and for the oil-importing developing countries the corresponding figure was 34 percent. For the world as a whole, it was about 56 percent.[9] Between 1970–72 and 1976–78, seismic effort[10] in the OIDCs actually declined by 8 percent and for the non-OPEC LDC oil exporters by 18 percent. On the other hand, for the industrial countries seismic effort rose by 46 percent.[11] The World Bank report also states that only about 14 percent of the estimated ultimately recoverable oil reserves of the OIDCs have been proved.[12]

In the OPEC countries drilling between 1970–72 and 1976–78 increased by only 5.3 percent and seismic effort declined by 3 percent. Moreover, the Middle East accounted for only 1 percent of all exploration efforts in 1980. The interest of many OPEC countries has been directed to conservation of petroleum resources while limiting output in order to maximize price. Also, the international oil companies have had little incentive to explore in the OPEC countries. However, several of the OPEC countries, including Indonesia, Nigeria, and Algeria, have been producing close to their sustainable production, and their revenues have been insufficient to meet their demand for imports. A larger exploration effort in most of the OPEC countries would undoubtedly lead to the discovery of significantly larger reserves.[13]

In a recent study by two French petroleum specialists, J. Favre and H. LeLeuch, it was estimated that during the 1975–79 period, 45 percent of the seismic prospecting (in terms of party months) in the non-OPEC LDCs took place in the exporting countries; another 43 percent

in other countries producing oil; and 12 percent in the nonproducers. They also found that during the same period 41 percent of the exploratory wells drilled in non-OPEC LDCs were drilled in the exporting countries; 50 percent in the other producing countries; and only 8 percent in nonproducing countries. For the non-OPEC LDCs as a group they found that total seismic prospecting activity represented only 17.1 percent of that for all the market economies during the 1975–79 period, and that only 4.2 percent of the exploratory wells drilled during that period were in non-OPEC LDCs.[14]

A 1983 World Bank study showed that between 1974 and 1980, seismic activity in the OIDCs rose by 130 percent, but there was no increase in seismic activity in the non-oil-producing OIDCs. About two-thirds of the seismic activity in the OIDCs was in the oil-producing countries. During the same period the number of exploratory wells drilled in the OIDCs rose by 27 percent, but in the non-oil-producing countries exploratory drilling actually declined.[15] Exploration activity was highly concentrated, with over 90 percent of the wells being drilled in the OIDC producing countries; nearly 60 percent of this drilling was in three countries—Argentina, Brazil, and India.[16] In 1974 the number of wells drilled in the OIDCs represented only 3.6 percent of the world total, declining to 3.1 percent in 1980.

In 1980, 71 percent of the exploratory wells in the OIDCs were drilled by national oil companies such as Petrobras in Brazil, the Oil and Natural Gas Commission in India, and Yacimientos Petroliferos Fiscales (YPF) in Argentina. Of the remainder, 20 percent were drilled by the seven major multinational petroleum companies;[17] 4 percent by developed country national companies such as Elf Aquitaine of France and ENI of Italy; and the remaining 5 percent by foreign and domestic independents.[18]

[9]World Bank, *Global Energy Prospects*, Staff Working Paper No. 489 (Washington, D.C., August 1981), p. 40, table 13. See also J. Favre and H. LeLeuch, "Petroleum Exploration Trends in the Developing Countries," *Natural Resources Forum* vol. 5 (October 1981) pp. 327–346; and T. C. Lowinger, "Petroleum Production in Developing Countries: Problems and Prospects," *Journal of Energy and Development* vol. VII (Spring 1982) pp. 235–237.

[10]Exploration activity prior to actual drilling measured in "party months."

[11]*Global Energy Prospects*, p. 41, table 14.

[12]Ibid., p. 41.

[13]Ibid., p. 43.

[14]Favre and LeLeuch, "Petroleum Exploration Trends," p. 335.

[15]World Bank, *Energy Transition*, p. 38.

[16]Ibid.

[17]Shell, British Petroleum, Mobil, Texaco, Chevron, Exxon, and Gulf.

[18]World Bank, *Energy Transition*, p. 38.

Although this study is mainly concerned with the role of foreign private investment in petroleum industries of the non-OPEC LDCs, it should be kept in mind that they are currently responsible for less than 30 percent of the exploration activity in these countries.

Implications of the Petroleum Resource Potential and Exploration Activity in the OIDCs

The above summary of the petroleum reserve potential and the volume and trend of exploration activity raises questions about the potential oil output of the OIDCs that have an important bearing on petroleum development policy and incentives for petroleum investment in these countries. A major question is whether a large expansion, say a doubling, of exploration activity in the OIDCs would lead over time to a proportionate rate of increase in their oil reserves. A related question is whether an "adequate" exploration effort in the nearly 90 OIDCs where oil has not been discovered in commercial quantities will result in a substantial number of discoveries. A third question is whether the relatively low level of exploration activity in the OIDCs is primarily a consequence of poor geologic prospects or is related to domestic and foreign investor policies.

Most petroleum geologists argue that enough is known about the geologic structure of the world to be able to project the probability of very large basins where giant (over 0.5 billion barrels) and supergiant fields (over 5 billion barrels) are likely to be found. Major accumulations of oil are usually found in the early stages of exploration with the drilling of a few wells. When a discovery occurs, it tends to induce a high level of exploration activity in the area, so that intensive drilling activity tends to *follow* exploration success rather than lead to it. The point is made, therefore, that while a substantial increase in exploration activity in a number of countries is warranted, there is only a small likelihood that there are large unsuspected oil fields in countries where there has been little exploration activity. Petroleum geologists generally reject the argument that because the "density" of wells drilled in a number of OIDCs where no oil has been found is low, an increase in the number of wells drilled will yield a proportionate rise in discoveries.[19]

Although it cannot be concluded that discoveries will increase substantially in the non-OPEC LDCs simply as a consequence of a general expansion of drilling density, a selective expansion of drilling, as indicated by geologic conditions and by the results of past exploration activities, should result in a significant increase in reserves. However, these reserves are more likely to be found in smaller fields of less than 500 million barrels rather than in giant fields. Table 2-1 shows the non-OPEC LDCs that were producing crude oil in 1982. It is these countries, plus a few others in which oil has recently been discovered (but not produced), that are likely to supply the bulk of the increase in output from the non-OPEC LDCs during the rest of this century.

While few giant fields are likely to be found in the OIDCs, the potential for finding small fields (under 50 million barrels) appears to be favorable. The U.S. Department of the Treasury study referred to above raises the question of whether small fields are likely to be attractive to petroleum investors since capital costs per barrel tend to rise as the field size declines and consequently the rate of return on an investment is much lower for small fields. The Treasury study shows, however, that "in the absence of royalties or taxes an average field of only 25 million barrels would provide a real internal rate of return of nearly 15 percent and that consequently private companies have incentives to develop smaller fields in OIDCs as long as government terms and geologic prospects are reasonably favorable."[20]

---

[19]For a review of the exploration literature and a discussion of this problem, see Joy Dunkerley, William Ramsay, Lincoln Gordon, and Elizabeth Celcelski, *Energy Strategies for Developing Nations* (Baltimore, Md., Johns Hopkins University Press for Resources for the Future, 1981) pp. 128–134.

[20]Office of International Energy Policy, *World Bank Energy Lending*, pp. 8–9.

Prospects for Expanding Oil Output in the OIDCs

A 1983 World Bank staff report states that not only can OIDC oil production more than double between 1980 and 1995, but the number of OIDCs that were oil producers in 1980 (20 countries) can be increased substantially by 1990.[21] Oil has been discovered in five OIDCs that were not producing in 1980—Benin, Chad, Niger, Ivory Coast, and Sudan—and there are 15 other OIDCs where exploration is active and where prospects of discovery appear promising. For those OIDCs not producing in 1980, a 1980 World Bank report projects an output of between 0.7 mbpd and 1.5 mbpd, depending upon the level of exploration and development activities in those countries.[22]

What are the conditions for achieving the warranted level of exploration activity in the OIDCs, and in the nonproducing OIDCs in particular? First, it should be noted that while GOEs have been responsible for nearly three-fourths of the exploratory wells drilled in the OIDCs over the 1974–80 period, some 70 percent of the wells drilled by the GOEs have been drilled in Argentina, Brazil, and India. In 1980 the number of exploration wells drilled by GOEs was only 38 percent above that in 1972. For GOEs other than those in Argentina, Brazil, and India, the number of wells drilled in 1980 was only 15 percent above those drilled in 1972. By contrast, in 1980 the number of exploratory wells drilled by foreign companies and private domestic companies was 55 percent above the number drilled in 1972.[23] Given the multiple rise in oil prices since 1972, it seems clear that GOEs should have undertaken more exploration. An alternative would have been for GOEs or their governments to invite foreign petroleum companies to undertake more exploration and development, but several of them, e.g., Brazil and India, did not do so until the late 1970s.

Excluding exploratory wells drilled by GOEs in Argentina, Brazil, and India, foreign companies and domestic independents were responsible for over half of the rest of the exploratory wells drilled in the OIDCs during the 1974–80 period. Moreover, most of the exploration activity in the nonproducing OIDCs, including seismic exploration, has been undertaken either by foreign companies or by GOEs with the assistance of the World Bank.[24] With few exceptions, OIDCs do not have GOEs with sufficient technical knowledge and experience to undertake an adequate exploration program. This is particularly true in the case of deep offshore drilling where, in a number of OIDCs, the best prospects for discovering reserves exist. For most OIDCs, a substantial rise in petroleum production, on the order of the potential rise estimated by the World Bank between 1980 and 1995, will depend upon the growth of foreign exploration and development activities in these countries.

The 1980 World Bank study concludes that unless the level of exploration activity in the nonproducing OIDCs can be raised in the near future, there is little chance of their producing a substantial amount of oil in the coming decade, given the time needed to mount an exploratory campaign and develop discoveries to the point of starting commercial production. The World Bank study further concludes that to attract risk capital toward nonproducing countries whose potential the oil companies have evaluated as small, and which they consider politically unstable or which have legislative or contractual provisions that deter foreign preservation, "more purposive actions by governments and international institutions may be necessary."[25]

The Petroleum Potential of Individual Non-OPEC LDCs

In 1980, of some 130 non-OPEC LDCs, 14 were net oil exporters. These included Angola,

[21]World Bank, *Energy Transition*, p. 13.

[22]World Bank, *Energy in Developing Countries*, pp. 15–16.

[23]The World Bank's exploration program in nonproducing OIDCs is discussed in chapter 12.

[24]World Bank, *Energy Transition*, p. 38.

[25]World Bank, *Energy in Developing Countries*, p. 16.

Bahrain, Brunei, Cameroon, Congo, Egypt, Malaysia, Mexico, Oman, Peru, Syria, Trinidad and Tobago, Tunisia, and Zaire.[26] The net export position of a few of these countries, including Peru, Syria, and Zaire, is precarious because their consumption is rising relative to their production, and without a significant increase in their reserves, they are likely to become net oil importers during the 1980s. Argentina, Bolivia, and Burma are in near balance; and Chile, Colombia, Ghana, and India produce 25 percent or more of their oil consumption. Another 11 countries have some commercial production, including Barbados, Brazil, Guatemala, Israel, Ivory Coast, Morocco, Pakistan, Philippines, Taiwan, Thailand, and Turkey. This leaves nearly 100 non-OPEC LDCs that do not have commercial production, or at least production has not been recorded in the data sources for 1980. Many of these countries are small island nations, but they include a number of African countries such as Botswana, Ethiopia, Kenya, Liberia, Madagascar, Mozambique, Senegal, Sierra Leone, Somalia, Sudan, Tanzania, Togo and Zambia; Guyana, Paraguay, Uruguay and all the Central American countries, except Guatemala, in Latin America; most of the Caribbean countries; North and South Korea and Sri Lanka in Asia; and Jordan, Lebanon, and Yemen in the Middle East. It should be mentioned, however, that in a few of these countries where exploration has taken place, oil has been discovered.[27]

Despite the fact that there has been some exploration, including exploratory drilling, in nearly all of the non-oil-producing countries mentioned above, no commercial reserves have been found in the bulk of them and industry geologists regard most of them as having a poor petroleum potential. However, there have been surprises in petroleum exploration in the past and they may occur in the future; it is likely that petroleum reserves will be discovered in some OIDCs where an adequate exploratory program is undertaken.

Geologic potential is, of course, not the only factor in determining a country's petroleum outlook. Differences in exploratory effort among countries have been greatly influenced by political and economic factors. Many LDCs made little effort to find or expand their petroleum reserves until after the severalfold rise in petroleum prices in 1974. In some countries, such as Brazil, barriers to exploration by international petroleum companies were not relaxed until recently, and in many countries substantial barriers remain. Brief reviews of both the geologic potential and the economic and political conditions for petroleum activity are given in the following paragraphs for several non-OPEC LDCs.[28] These reviews illustrate the problems frequently encountered in developing countries.

Argentina

In 1982 Argentina's proved reserves were 2.59 billion barrels and production was 483,000 bpd (table 2-1). Both proved reserves and production declined in 1982 from the previous year. Argentina's output relies heavily on secondary recovery from older wells and its 1982 proved reserves were sufficient for less than fifteen years of production at the 1982 output. Output is almost sure to decline in the near future unless there is a substantial increase in reserves.[29]

Argentina's ultimately recoverable reserves are a subject of considerable dispute among petroleum geologists, with the more optimistic estimates depending heavily on what may be discovered offshore.[30] Onshore exploration has been fairly intensive over several decades, while offshore exploration undertaken by international oil companies is of recent origin. The first discovery of offshore oil was made in Tierra del

[26]Excludes Soviet bloc countries and China. China was also a small exporter in 1980.

[27]With the completion of a pipeline, Sudan expects to be producing 100,000 bpd by 1985.

[28]Judgments regarding the recoverable oil potential for individual countries examined are based in part on resource estimates by Nehring in *Giant Oil Fields* and Arthur G. Warner, *The 1985 Oil Production of Twenty-One Oil Producing Non-OPEC Countries* (Washington, D.C., U.S. Department of Energy, March 1979). In some cases the judgments are based on conversations by the author with petroleum geologists.

[29]Argentina's 1982 annual production was about 176 million barrels.

[30]Nehring estimates Argentina's recoverable oil resources at 5.2 billion barrels. See Nehring, *Giant Oil Fields*, p. 32.

Fuego by a Shell group in 1981.

The average production of Argentine wells is only about 70 bpd. A large number of wells must be drilled each year as the older wells become depleted. By 1978, some 26,000 wells had been drilled since production was initiated, of which 6,000 were in operation. In 1980 Argentina was producing about 92 percent of its petroleum requirements, but over half the output was from secondary recovery. The bulk of Argentina's crude output is produced by the government's petroleum enterprise, Yacimientos Petroliferos Fiscales (YPF), and by private contractors operating in areas with proved reserves. A small amount is being produced under old concession contracts held by local private companies. Until 1980, no contracts for grass roots or risk exploration by private companies had been let since the 1960s. Following the passage of the Risk Contract Law of 1979, contracts were let to private petroleum companies for exploration and development in zones (both onshore and offshore) where little or no exploration activity had taken place. However, only a few risk contracts had been negotiated by the end of 1982, and it will be several years before the exploration programs are fully operational. Argentina's financial difficulties, together with the Falkland Islands war, have reportedly retarded the country's exploration and development activities.[31]

Bolivia

Bolivia's production of crude oil and condensate peaked at 47,300 bpd in 1973 and exports averaged 32,500 bpd. At that time there were 16 companies or joint ventures operating in the country; by 1981 there were only two, Occidental Petroleum and Tesoro Petroleum, both of which were operating under production-sharing contracts with the state petroleum company, Yacimientos Petroliferos Fiscales Bolivianos (YPFB). By 1982 total crude and condensate production had declined to 24,000 bpd, and proved reserves were estimated to be only 180,000

barrels—about seven years of output. Most of the recent discoveries have been gas from which condensate is produced. In 1980 the World Bank and the Inter-American Development Bank approved loans totaling $32 million to finance the drilling of 12 appraisal wells by YPFB.[32] Bolivia's potential for future production has been estimated at 100,000 to 115,000 bpd, but this is not expected to be achieved in the near future and without a substantial increase in exploration effort.[33] Political unrest and the absence of a favorable foreign investment climate have contributed to Bolivia's poor performance. Its recoverable oil resources are estimated at about 400 million barrels.[34]

Brazil

Brazil's 1982 output was 250,000 bpd, 17 percent above that of the previous year and 37 percent above the 1980 level. Brazil's proved reserves also rose substantially, from 1.33 billion barrels in 1981 to 1.75 billion barrels in 1982. Brazil is one of the most active drilling areas outside the United States, with 53 rigs operating onshore and 29 offshore during August 1981. Although the decline in onshore production was reversed in 1981, the more favorable source is believed to be in deepwater areas.[35] Until late 1981, all production and significant discoveries were made by the state petroleum agency, Petrobras. Brazil has opened large areas for risk contracts with private petroleum companies and over 100 contracts had been signed by the end of 1981, with a large number of international companies represented. Despite the drilling of 52 wells by private contractors, only one discovery had been reported—that by the Shell group in January 1982. Evidently the international companies believe Brazil has a substantial potential for new discoveries and the more recent Brazilian risk contracts are regarded as relatively attractive.

[31]"Argentina Eyes More Private Oil Work," *Oil and Gas Journal* (December 6, 1982) pp. 121–122.

[32]Frank E. Niering, Jr., "Bolivia: Hopes for Renewing Oil Exports," *Petroleum Economist* (September 1981) pp. 381–383.
[33]Warner, *1985 Oil Production*, pp. 45–46.
[34]Nehring, *Giant Oil Fields*, p. 33.
[35]Warner, *1985 Oil Production*, p. 37.

Brunei

Brunei, a small enclave on the west coast of Borneo, has proved reserves of 1.24 billion barrels (mostly offshore) and was producing at a rate of 155,000 bpd in 1982, a sharp reduction from 1979 when output was 255,000 bpd. Production is in the hands of a 50-50 joint venture involving Shell Petroleum (BSP) and the Brunei government.[36] Prospects for raising output from existing Brunei reserves are good, but are doubtful for a substantial rise in proved reserves, estimates of which have been declining in recent years. Exploration drilling also has been low, but in 1982 BSP announced plans for an expansion.[37]

Chile

Chile's crude production based on onshore wells declined from a peak of 17.3 million barrels in 1973 to 12.7 million barrels in 1977, with a further decline in 1978. However, in 1979 four offshore wells came on-stream and production doubled between 1978 and 1980, with a further increase to an estimated 41,000 bpd in 1982. All of the increase came from offshore production while onshore output continued to decline. In 1981 all the crude was produced by the state enterprise, Empresa Nacional de Petroleo (Enap). In 1977 Enap negotiated a thirty-five-year risk contract with a joint venture involving Arco and Amerada Hess for offshore exploration and development. In 1981 the joint venture drilled three wells, but all were unsuccessful.[38]

Chile's estimated proved reserves in 1982 were less than 0.8 billion barrels (see table 2-1). It is unlikely to be a large producer but may achieve self-sufficiency in petroleum by 1990, depending upon the success of the offshore exploration by Enap and the international companies. Chile expects to negotiate risk contracts for additional west coast marine areas. Although contract terms have been relatively generous, substantial

interest on the part of international companies is unlikely to occur until significant discoveries are made in the contract areas. Chile's recoverable oil resources have been estimated at 400 million barrels, about the same as the estimate for Boliva.

Colombia

Oil production in Colombia peaked in 1970 at around 219,000 bpd, fell steadily to 124,200 bpd in 1979, and rose to an estimated 140,000 bpd in 1982. However, production from existing fields is expected to decline rapidly and a continuation of the rise in output will depend heavily on output from new fields, several of which have been discovered recently. Estimates of reserves range from about 540 million barrels in 1982 by the *Oil and Gas Journal* (see table 2-1) to 800 million by Frank Niering.[39] Seismic activity and exploratory drilling were substantial during the 1980–82 period and involved both the government's petroleum enterprise, Ecopetrol, and several international oil companies, including Exxon, Elf Aquitaine, Texaco, and Houston Oil. All of these companies signed new contracts of association under the 1976 model contract, which is more liberal to the contractors than previous arrangements. The decline in Colombia's production during the 1970s was due in considerable measure to unattractive contract arrangements with foreign companies and a low level of exploratory activity on the part of both private oil companies and Ecopetrol. Colombia's recoverable oil resources have been estimated by Niering at 2.6 billion barrels, about half that estimated for Argentina.

Egypt

Egypt's petroleum production has increased rapidly over the past several years, from 415,000 bpd in 1977 to 667,000 bpd in 1982.[40] Egypt's proved reserves were an estimated 3.3 billion barrels in 1982. The recent rise in output is

[36]"Brunei: The 'Gulf State' of the Far East," *Petroleum Economist* (June 1978) pp. 248–249.
[37]*Petroleum News* (January 1982) p. 10.
[38]*Petroleum Economist* (August 1981) p. 360; see also, "Chile Beckons Foreign Firms for Contract Areas," *Oil and Gas Journal* (October 25, 1982) pp. 92–93.
[39]Frank E. Niering, Jr., "Colombia: Upturn in Oil Output and Drilling," *Petroleum Economist* (February 1982) p. 50.
[40]A portion of this increase may be attributed to the transfer of the Sinai field from Israel.

principally due to the discovery and development of five giant fields in the Suez basin. Thirteen oil finds were made during 1980 and 15 in 1981 and early 1982. On this basis, Egypt's oil minister believes that Egypt will reach 1 million bpd by the mid-1980s.

Egypt's success has been attributed in considerable measure to its policy of attracting foreign oil companies during the 1970s. Since 1973, 75 exploration concessions have been granted to 49 companies, with almost all the major companies represented. In contrast, in the decade from 1963 to 1973, Egypt had only four exploration agreements with three companies.[41] In comparison with OPEC producers, Egypt's terms for its production-sharing concessions are relatively generous. On the award of an exploration concession, a company must commit a specified sum—$10 to $25 million—to explore over a period of four to eight years, and must pay a signature bonus of from $1 to $3 million. If a commercial find is made, a 50-50 joint-venture company is formed with the oil company and the state enterprise, Egyptian General Petroleum Corporation (Egpc). The company is allowed a share of production, usually 30 percent, for cost recovery, with the remainder being shared between the government and the oil company. The company typically receives 15 to 20 percent of the profit oil with a lower share at high production rates, but substantial production bonuses must be paid as output builds up. Egpc pays all taxes and royalties, but contributes nothing toward exploration and production costs.[42]

Guatemala

Petroleum was discovered in Guatemala in 1974; production began in 1979 at about 2,000 bpd and rose to 5,000 bpd in 1980 and 1981 and to 6,000 bpd in 1982. There was relatively little interest in Guatemala until large discoveries were made in Mexico's Reforma area near Guatemala's northern border in the late 1970s. In 1978

contracts were signed by a consortium headed by Getty Oil, Hispanoil, and Texaco/Amoco, and by a Texaco consortium in 1982.[43] Several discoveries were announced in 1981 and 1982, but Guatemala's proved reserves were estimated at only 50 million barrels in 1982. As will be discussed in chapter 7, the interest of foreign companies in Guatemala has been limited by unfavorable contract terms. Guatemala's recoverable oil resources were estimated by Niering at 300 million barrels as of 1975, but recent discoveries across the Mexican border have probably raised estimates of Guatemala's petroleum potential.

India

India's estimated proved reserves as of 1982 were 3.4 billion barrels, up from 2.7 billion barrels in 1981, and output was 384,000 bpd in 1982 (table 2-1) contrasted with 284,000 bpd in 1981. The recent rise in India's reserves and production has occurred mainly in the offshore fields. India's state oil companies dominate petroleum activity, and prior to 1981 only one foreign contractor, Burmah Oil of the United Kingdom (in a joint venture with the government), was active in drilling. Until the mid-1970s, the Indian government sought to rely entirely on state oil enterprises, but during 1974-75 the government signed production-sharing contracts with three foreign companies, all of which withdrew after a limited amount of offshore drilling. A joint venture between an Indian GOE and Burmah Oil for offshore exploration was formed in 1978. The terms of India's contracts were generally unattractive, in part because the contractor was required to sell its cost and profit oil to the government at relatively low prices, and in part because companies were required to surrender concessions five years after the start of production. This was much too short a period for companies to recover their costs with a profit.

Faced with a mounting oil import bill and limited financial and technical resources, the

[41]"Egypt on the Threshold of Oil Boom," *Wall Street Journal* (February 5, 1982) p. 31.

[42]Martin Quinlan, "Sadat's Oil Success," *Petroleum Economist* (September 1981) pp. 522-524.

[43]*Oil and Gas Journal* (August 9, 1982) p. 71.

government again decided in August 1980 to invite foreign oil companies to explore for oil and gas on the basis of a production-sharing formula. Protracted discussions were carried on with a number of international companies, but contracts could not be negotiated under terms offered by the government, which were similar to those provided in the three production-sharing contracts signed earlier. Finally the government relented on the price issue and, although foreign companies will not be allowed to export oil until India becomes self-sufficient, the petroleum minister announced that India would buy the companies' oil share at international prices. It is also reported that India extended the terms of the contracts. In December 1981 a contract was signed with Chevron Overseas Petroleum for an offshore block. India has had little success in attracting exploration by foreign petroleum companies and Chevron was the only non-Indian company conducting exploration early in 1983. India's contracts are reported to be unattractive to the industry. Many of the blocks being offered for contracts are in very deep water where drilling costs are high.[44]

The Indian government has received substantial financing from the World Bank and other international institutions for operating the state oil enterprises, which have been among the most successful of the GOEs operating in the LDCs.

Ivory Coast

Ivory Coast has only recently become a petroleum producer, with production initiated at about 7,000 bpd in 1980 in a field discovered by Esso in 1975. In 1980 the offshore Espoir field was discovered, with reserves estimated as high as 8 billion barrels, and production began in August 1982 at a rate of 9,000 bpd from four wells.[45] Press reports indicate that by 1985 the Espoir field alone may be producing up to 450,000 bpd. The field was discovered and is being exploited by a consortium consisting of Phillips Petroleum and Sedco of the United States, Agip of Italy, and Petroci (Ivory Coast's national oil company), with Phillips as the operator. Ivory Coast output may soon equal that of other West Coast African producers, such as Angola, which produced 122,000 bpd in 1982; Cameroon, 109,000 bpd in 1982; the Congo, 87,000 bpd in 1982; and Gabon, 130,000 bpd in 1982.

The Ivory Coast discoveries were evidently something of a surprise to many geologists since there is no record of that country's oil potential being rated in the literature prior to 1980. The rapid success in both the discovery and development of Ivory Coast petroleum may be attributed in considerable measure to the attractive contracts offered to international petroleum companies, strong government cooperation, and a generally favorable economic climate.

Malaysia

Malaysia's proved reserves in 1982 were an estimated 3.3 billion barrels, substantially above the 1.8 billion barrels in recoverable oil resources estimated by Nehring as of December 31, 1975.[46] Malaysia began producing at the rate of 100,000 bpd in 1975, with output rising to 288,000 bpd in 1980, but declining to 259,000 bpd in 1981 as a consequence of the Malaysian government's policy of curtailing output of existing fields. This policy was evidently reversed in 1982 when production rose to an estimated 306,000 bpd. There was also a sharp rise in proved reserves, indicating a successful exploration program. For the past several years the Malaysian government and its state petroleum enterprise, Petronas, have had serious disputes with some of the foreign petroleum companies that negotiated contracts, but recently it has been reported that contract terms have been improved (see chapter 8).[47] The outlook for a continued rise in Malaysian production and exports appears to be favorable given the 30 to 1 reserve–output ratio and annual additions to reserves.

[44]"India," *Petroleum News* (January 1982) pp. 17–18; and "Blocks Off India Draw Little Interest," *Oil and Gas Journal* (February 14, 1983) p. 62.
[45]"Ivory Coast Early Production System Starts Up," *Oil and Gas Journal* (September 13, 1982) p. 32.

[46]Nehring, *Giant Oil Fields*, p. 33.
[47]"Petronas Improves Cost Recovery Provisions," *Oil and Gas Journal* (November 29, 1982) p. 48.

Peru

Peru's oil production took a sharp jump in 1978 with the coming on stream of Occidental's discoveries in the eastern jungle, reaching 195,000 bpd in 1979 but declining to 184,000 bpd by 1981, partly as a consequence of the depletion of old wells. Peru's 1982 production is estimated at 198,000 bpd, and there was a small rise in estimated reserves to 840 million barrels.[48] New contracts have been awarded to international petroleum companies for both offshore areas and blocks in the eastern jungle, and Petroperu (the state oil enterprise) has been quite active in its drilling program. Evidently Peru's petroleum potential is believed to be substantially greater than was envisaged in 1975 when Nehring estimated its ultimately recoverable oil resources at only 1.7 billion barrels.

The Peruvian government has had serious disputes with foreign petroleum contractors, including a forced renegotiation of contracts held by Belco Petroleum and Occidental Petroleum (the two largest private producers) in 1980. However, contracts with Shell and Superior Oil signed in 1981 provided somewhat more generous terms for the companies and there is increased interest on the part of other companies in negotiating contracts. Peru has been exporting about 50,000 bpd annually, but its reserves must be increased substantially if the country is to continue to be a net exporter during the remainder of this decade.

The Philippines

Petroleum exploration and development in the Philippines, which began in the 1950s, has been disappointing. A large number of international companies have held concession or service contracts, and hundreds of wells have been drilled, with few discoveries. The reservoirs that have been discovered, entirely offshore, have been

small, and output from the few producing wells that have come on stream has declined rapidly. Production peaked at 26,000 bpd in 1979, declining to only 2,000 bpd in 1981, and increasing to 7,000 bpd in 1982. Proved reserves for 1982 were estimated at 36 million barrels by the *Oil and Gas Journal* (see table 2-1), and a recent estimate in the *Petroleum Economist* put reserves at "no more than 90 million barrels."[49] The outlook for the country's becoming self-sufficient in petroleum—imports were 76 million barrels in 1981—is unfavorable.

Summary and Conclusions

In 1982 the non-OPEC LDCs produced 17 percent of the crude petroleum output of the market economies and held about 14 percent of the proved reserves. However, of the more than 130 non-OPEC LDCs, 14 exporting countries produced over 80 percent of the crude petroleum and held about 80 percent of the reserves, with Mexico alone accounting for over half the reserves. Of the remaining non-OPEC LDCs, three produced over 90 percent of their consumption requirements, another four produced 25 percent or more of their oil consumption, while another 11 countries had some commercial production. Nearly 100 non-OPEC LDCs did not have commercial production in 1982.

The greatest part of the seismic prospecting and exploratory drilling during the 1970s took place in developed countries and in OPEC members, and most of the remainder took place in the non-OPEC LDCs that were already producers. It cannot be concluded, however, that the relatively low share of world oil reserves in the non-OPEC LDCs simply reflects the share of these countries in world exploration activities, or that most of the nonproducing LDCs would have been significant producers if exploration activities in these countries had been increased severalfold. This is not to say that oil-importing countries with considerable petroleum potential, such as Brazil, would not have been much closer

[48] Petroperu estimated Peru's reserves early in 1982 at 900 million barrels, somewhat larger than the 1982 proved reserve estimate given by the *Oil and Gas Journal* December 27, 1982, pp. 78–79 (see table 2-1). See "Peru: Drive to Boost Production," *Petroleum Economist* (March 1982) p. 109.

[49] G. V. Hough, "Philippines: Exploration Hopes Offshore Palawan," *Petroleum Economist* (December 1982) pp. 507–508.

to petroleum self-sufficiency if their exploration efforts had been substantially greater during the 1970s and earlier, or that significant discoveries would not have been made in some of the present nonproducers if their exploration activity had been greater. Significant oil discoveries have been made in several countries that were not producers in 1980, and prospects for discoveries appear favorable on the basis of seismic and other exploration activity in a number of others. An increase in exploration activity appears warranted in the OIDCs.

Although more than half the exploratory wells drilled in the OIDCs over the 1974–80 period were drilled by GOEs, most of this drilling has been undertaken by only a few GOEs with considerable technical capacity, experience, and financial resources. This suggests that most OIDCs, and particularly the non-oil producers, will need to depend heavily on foreign petroleum companies for a major expansion of their exploration activities. Some of the countries where petroleum operations have been almost exclusively in the hands of GOEs in the past, e.g., Brazil and India, are now actively seeking exploration by foreign companies.

Among the countries discussed in this section, those with the greatest potential for substantial expansion of petroleum production include Brazil, Egypt, India, Ivory Coast, Malaysia, and Peru. International petroleum companies are operating in all of these countries and, except for Brazil and India, most or all of the recent discoveries have been made by the foreign private companies. Argentina's future production will depend heavily on offshore discoveries since the mainland has been extensively explored over many decades, but offshore exploration by international oil companies has been initiated only recently. The potential for the Philippines seems quite limited in view of the lack of success in discovering fields capable of significant production, and much the same can be said of Chile, whose future petroleum output also depends heavily on significant offshore discoveries. Colombia, which was at one time a petroleum exporter but whose progress was interrupted by unattractive contractual terms for foreign producers, has only recently expanded its exploration effort. Bolivia's output has been continuously declining since 1974 as a consequence of a combination of poor prospects and an unfavorable political and economic climate. Guatemala's potential cannot be determined without considerably more exploration, which has been limited by relatively unfavorable contract terms offered to foreign petroleum companies.

3

Evolution of Petroleum Contracts Between Host Governments and the International Petroleum Companies

Origin of Oil Concessions

The exploration and production of petroleum in developing countries by international oil companies date from the latter part of the nineteenth century and the operations of the Royal Dutch Company (later Royal Dutch Shell) in the Dutch East Indies (Indonesia); these were followed by a British group (later the Anglo-Persian Oil Company, predecessor of British Petroleum), which first obtained a concession in Iran in 1901. American firms began acquiring oil properties in Mexico in the early 1900s.[1] Although the British government took a strong interest in petroleum prior to World War I and acquired a 53 percent interest in the Anglo-Persian Oil Company in May 1914, access to petroleum became a major national security objective of the world powers after the war, and national governments gave strong support to the international oil companies. British political hegemony in the Middle East and Asia provided backing for British companies in that area, while U.S. political influ-

ence in Latin America provided a favorable environment for U.S. petroleum companies in that region. Royal Dutch Shell, which was a Dutch–British joint venture, dominated petroleum production in the Far East for many years. By the late 1920s, seven American, British, and Dutch–British oil companies (the Seven Sisters) controlled most of the oil produced in Latin America, the Middle East, and Far East. They were Exxon (formerly Standard Oil Company, N.J.); Mobil (formerly Socony-Vacuum Oil Company); Gulf Oil Corporation; Texaco; Standard Oil Company of California (Socal); British Petroleum Company Ltd. (BP); and Royal Dutch Petroleum Company and Shell Transport and Trading (Shell). An eighth important company with early interests in the Middle East and subsequent widespread international operations is Compagnie Francaise des Petroles (CFP) and its various subsidiaries and affiliates.

The discovery of oil in Persia (Iran) stimulated exploration in Mesopotamia, but following the war, British, French, and Dutch interests were consolidated into the Turkish Petroleum Company, and the British government, which

---

[1]Neil H. Jacoby, *Multinational Oil* (New York, Macmillan, 1974) pp. 26–27.

governed the area under a League of Nations Mandate, sought to keep American companies from obtaining concessions. U.S. government protests against the exclusion of American interests eventually led to the creation of the Iraq Petroleum Company (IPC) in which American interests were accorded about a quarter share. In the final configuration, each of the following held 23.75 percent of the shares: (a) Anglo-Iranian Oil Company, (b) Royal-Dutch Shell, (c) Cia Francaise des Petroles, and (d) the Near-East Development Corporation (owned jointly by Standard Oil Company, N.J. and Socony-Vacuum). The remaining 5 percent was owned by C. S. Gulbenkian, one of the owners and promoters of the old Turkish Petroleum Company. In 1928 these companies negotiated the so-called "Red Line Agreement," which provided that each participant in IPC would refrain from taking separate action in an area that included most of the old Ottoman Empire. For a time the Red Line Agreement limited the expansion of Jersey and Socony-Vacuum, but eventually four American companies—Standard Oil of California (Socal) (which obtained the original concession), Jersey, Texaco and Mobil—organized the Arabian-American Oil Company (Aramco) which held a concession over much of eastern Saudi Arabia. Anglo-Iranian and Gulf Exploration each held a 50 percent interest in the Kuwait Oil Company, while Socal and Texaco each held a 50 percent interest in the Bahrain Petroleum Company.[2]

Elements of Traditional Concession Agreements

Although the early concession contracts differed from country to country, most of them had the following elements:

1. A definition of the concession area within which the company was given the right to carry on exploration and oil development. (The concession area usually covered thousands of square miles, sometimes a major part of an entire country.[3])

2. A minimum amount of drilling that must be done over a period of time before oil is found in commercial quantities.

3. The duration of the concession, usually from sixty to seventy-five years.

4. The financial obligations of the company, which usually took the form of lump-sum payments, an annual rental, and royalties on each barrel of crude produced.

5. A provision for supplying the oil requirements of the local economy, either free of charge or at prices below those prevailing in world markets.

6. Installation rights and the right of eminent domain.

7. Certain special rights such as freedom from taxation other than that fixed in the contract, and freedom from government controls over the conditions of production and marketing.

In countries where the rights to minerals in the subsoil were not owned by the government[4] and where the government was not the owner of the land, oil lands were either leased or purchased from the private landowners by the petroleum company. For example, in 1913 Standard Oil Company, N.J. purchased controlling interest in the London and Pacific Petroleum Company of Peru, which gave Standard Oil ownership of an oil field, the title to which had been originally granted to a Peruvian in 1826 under an agreement which provided for an extremely light tax on the property. In 1914 the Peruvian government demanded tax payments on the property in accordance with regular Peruvian law. This

[2]For a history of the concessions in the Persian Gulf area, see Raymond F. Mikesell and Hollis B. Chenery, *Arabian Oil: America's Stake in the Middle East* (Chapel Hill, N.C., University of North Carolina Press, 1949) chapter 4.

[3]Aramco's concession in Saudi Arabia covered 440,000 square miles and IPC's concessions, together with those of its subsidiaries, covered nearly the entire territory of Iraq.

[4]Most developing countries follow the practice of vesting ownership of minerals in the subsoil in the state, but some have not done so until recent years. Brazil's constitution did not vest ownership by the government in minerals in the subsoil until 1934.

led to a dispute between the Peruvian government and Standard Oil's subsidiary, International Petroleum Company (IPC), which was not completely resolved until the military government under President Velasco expropriated IPC in January 1969.[5] In other countries foreign companies acquired mining and petroleum rights under the governments' general mineral codes in accordance with the same terms that applied to domestic holders of mineral rights. However, as time went on, most countries established a special contractual and fiscal framework for petroleum that differed from provisions relating to nonfuel minerals in the mining codes, so that it became necessary to negotiate special agreements with the governments to explore for and develop petroleum.

Pricing Under Concession Contracts

Under the traditional concession agreement prior to the 1950s, the oil companies determined the prices of the crude they sold to their downstream affiliates. Since royalties were based on quantities exported and income taxes were not generally employed until the 1950s, the host governments had no direct interest in prices. The nearest approach to a free market price was the spot market on the Texas Gulf, but this was a very thin market. Most of the oil entering the United States was transferred from producing affiliates of international companies operating abroad to their affiliates in the United States. In addition to world-wide collusive arrangements by the international companies for controlling oil prices, U.S. producers' prices were from time to time regulated by the U.S. government. Although the cost of Venezuelan and Middle Eastern oil was less than the cost of U.S.-produced oil, oil imported from abroad by the major companies tended to be sold at U.S. prices. In times of surplus there were often discounts from U.S. prices on foreign oil. Much of the Middle East oil, and some of the Venezuelan and Texan

oil as well, went to Western Europe, and an effort was made to equalize crude prices in England and on the Continent whether the oil came from the Middle East or the Caribbean. This led to a system of "posted" prices, established by companies producing in Middle Eastern countries, which were related to Texas Gulf prices but were set at levels which, after adding transportation, would compete with Caribbean oil in Western Europe and at times in the United States as well. As the cheaper Middle Eastern oil output expanded, prices were set at levels which enabled this oil to compete further west.

The system of "posted" prices was initiated in 1950 at about the time that Middle Eastern governments were introducing income taxes as a replacement for, or an addition to, royalties based on output. The "posted" prices determined the price used for calculating both income tax obligations and royalties based on the value of output. Hence, the Middle Eastern governments began to take a keen interest in prices since they determined their revenues. Venezuela had introduced income taxes in 1942 and the principle of a 50-50 split in profits between the oil companies and the Venezuelan government became the rule in 1948.[6] The Middle Eastern governments also adopted the 50-50 split rule for division of profits in the early 1950s.[7] Since taxable income was based on the posted prices established by the companies, a cut in Middle Eastern posted prices made by the companies in 1960 was bitterly resented by Middle Eastern producing countries, whose representatives argued that the major companies had unduly depressed prices of Middle Eastern oil. In Venezuela there was a long dispute between the oil companies and the government over pricing of petroleum for tax purposes.[8]

[5]Charles T. Goodsell, *American Corporations and Peruvian Politics* (Cambridge, Mass., Harvard University Press, 1974) pp. 53–54.

[6]See Gertrud G. Edwards, "Foreign Petroleum Companies and the State in Venezuela," Raymond F. Mikesell, ed. *Foreign Investment in the Petroleum and Mineral Industries* (Baltimore, Md., Johns Hopkins University Press for Resources for the Future, 1971) pp. 105–107.

[7]For a discussion of developments under Aramco's concession in Saudi Arabia, see Donald A. Wells, "Aramco: The Evolution of an Oil Concession," ibid., pp. 216–236.

[8]For a discussion of the dispute, see Edwards, "Foreign Petroleum Companies," pp. 115–116.

The disputes over prices between oil companies and governments of the oil-exporting countries in the Middle East and Venezuela constituted a major reason for the formation of the Organization of Petroleum Exporting Countries (OPEC) in 1960.[9] Initially OPEC sought to influence the posted prices, which became the basis of royalties and income taxes even though world prices and, therefore, transfer prices were lower in periods of oil surplus. However, it was not until the early 1970s when surplus oil-producing capacity disappeared that OPEC achieved the power to determine world oil prices. Since 1974 the governments of OPEC countries have determined prices for purposes of royalty and income taxes, while in the non-OPEC countries, prices were either based on OPEC prices or in some cases the concession holders or other types of contractors were required to sell the oil they produced to a national petroleum agency at prices set by the government. Where private producers had complete freedom to sell their oil on world markets, they were sometimes able to obtain prices in the spot market that were higher than the OPEC reference prices.

The Demise of the Traditional Concession Agreement

Many of the early concessions in the Middle East and Asia were negotiated while the countries were under the political control of European powers. As these countries became independent, they began to exercise greater control over their mineral wealth with a view to maximizing state revenues from oil production and to carrying out national policies regarding the development of their resources. The governments of Latin American countries, most of which had been independent since the middle of the nineteenth century, began adopting national economic policies after World War I, and the foreign-owned oil companies were among the first objects of national control. The actions of governments took two forms: first, they demanded renegotiation of earlier concession contracts with petroleum companies; and second, they created government oil enterprises (GOEs) to carry out national petroleum policies. In many countries these state enterprises came to completely dominate the country's petroleum operations.

The first government oil agency, Yacimientos Petroliferos Fiscales (YPF), was organized by Argentina in 1922 and given the right to enter all phases of the oil business, from exploration to transportation and marketing.[10] It also received from the government certain oil lands known as state reserves which it operated in competition with private companies in Argentina.[11] State oil companies were subsequently organized in Brazil, Chile, Peru, and Venezuela, and in 1937 Mexico nationalized that country's petroleum industry, placing it entirely under the operation of Pemex, the state oil company. By the early 1960s, nearly all developing country oil producers had established their own state enterprises, some of which were given a monopoly over all petroleum-producing activities. However, many of the GOEs, including Argentina's YPF, Indonesia's Pertamina, and Peru's Petroperu, negotiated exploration and production contracts with the former private-concession holders. In the mid-1960s Saudi Arabia's state oil enterprise, Petromin, negotiated joint ventures with international petroleum companies to develop areas outside the Aramco concession area, and later acquired a minority equity interest (which was eventually increased to 100 percent) in Aramco.

An important factor in the demise of the traditional concession system was the reduced bargaining power of the older international petroleum companies (i.e., the Seven Sisters) in competing for sources of crude in the developing world.

---

[9]The members of OPEC are Algeria, Ecuador, Gabon, Indonesia, Iran, Iraq, Kuwait, Libya, Nigeria, Qatar, Saudi Arabia, United Arab Emirates, and Venezuela.

[10]For a discussion of the evolution of government oil enterprises, see UN Centre for Natural Resources, Energy and Transport, *State Petroleum Enterprises in Developing Countries* (New York, Pergamon Press, 1980) chapters 1–4.

[11]See Gertrud G. Edwards, "The Frondizi Contracts and Petroleum Self-Sufficiency in Argentina," in Mikesell, *Foreign Investment*, p. 161.

The number of petroleum companies seeking American production greatly expanded as independent companies (e.g., Cities Service, Continental, Phillips Petroleum, Standard of Indiana, and Union Oil) plus the national oil companies of France and Italy began to rival in size and financial resources the Seven Sisters, which had controlled the international petroleum market prior to World War II. The independent petroleum companies began negotiating agreements on terms much more advantageous to the host countries than were the terms of the concession contracts previously negotiated by the older companies. Competition for concessions enabled host governments to force changes in the older concession agreements and to introduce new forms of agreements, such as joint ventures. Concession contracts were continually amended to give the national government not only an increasing percentage of the profits in the form of income taxes and other revenue-sharing arrangements, but also greater control over production and marketing. This occurred progressively in Venezuela until the final nationalization of the Venezuelan petroleum producing industry in 1976.[12]

Another factor in the demise of the traditional concession agreement was the increased role of GOEs in both exploration and production and in various forms of joint operations under contracts with private petroleum companies. The change in the structure of the international oil industry following the 1973 oil embargo imposed by Middle East OPEC countries and the creation of the OPEC oil cartel greatly reduced the dependence of oil-producing countries on a handful of international petroleum companies for world-wide marketing. Consequently, the government oil enterprises began to take over a larger share of marketing. In some cases they negotiated marketing contracts directly with national oil companies of developed countries such as France and Italy.

By the end of the 1970s, most Middle Eastern OPEC countries, including Saudi Arabia, Kuwait, Iraq, and Iran, had fully nationalized their petroleum industries and depended upon the private oil companies mainly for technology (including skilled operators and managers) and for marketing a portion of their output. Given large proven reserves and substantial financial resources, they no longer required high-risk capital from private oil companies for exploration and development of their petroleum. By 1980, the share of "production equity" of international oil companies in Middle East crude output had declined to less than 10 percent.[13] In many cases the former concession owners continue to provide technical services for a fee and have arranged to purchase substantial volumes of crude oil from the national oil companies in the host countries, but the amount of these purchases often depends upon the prices charged by the national oil companies.

Companies still maintaining contracts with OPEC countries[14] have in some cases lost their power to determine the volume of oil they can sell from their own output, and the price that determines the taxes on that output is set by the government. The prices established by the government often are not competitive in the world market and in times of world surplus the companies sell only a portion of the oil they produce. This has been the situation in Libya and Nigeria, for example. Much of the bargaining between governments and the remaining firms that have contracts has to do with prices, taxes, and contract violations, which are quite common. Recently some firms have pulled out of OPEC countries, as did Exxon from Libya in 1981 and Mobil in 1982. Having achieved effective control over prices and production as well as exploration for crude oil in their countries, the major OPEC countries have sought to control downstream operations such as shipping, refining, and marketing in their own countries and abroad.

[12]For a discussion of the shift in control over production and marketing from the multinational companies to the OPEC national oil companies, see G. E. Hartshorn, "From Multinational to National Oil: The Structural Change," *Journal of Energy and Development* vol. 5, no. 2 (April 1980)

[13]*Middle East Oil*, Exxon Background Series (New York, September 1980), p. 29.
[14]The OPEC members in which there is some form of foreign equity participation in petroleum include Algeria, Ecuador, Gabon, Indonesia, Libya, Nigeria, and the United

The situation is quite different for oil-importing countries or for countries, such as Peru, that are only marginally petroleum-exporting countries and require substantial exploration to maintain production. For countries without large known reserves, continual exploration is required if production is to be maintained. Even for some substantial exporting countries, such as Indonesia (an OPEC member), there is need for continuous exploration. This is also the situation with relatively small oil-producing countries such as Ecuador and Gabon, while the larger OPEC oil producers generally are not seeking to attract foreign risk investment.

Most non-OPEC LDCs have established their own national petroleum enterprises, which carry on varying amounts of exploration, development, and production. With a few exceptions, such as Mexico and India, a substantial amount of exploration and development activity is carried out by international petroleum companies under various forms of contractual arrangements with the national petroleum enterprises. Since large oil fields have not been discovered in most of the non-OPEC LDCs and it is necessary to conduct high-risk and often very costly exploration programs in offshore areas and in jungles and mountainous regions, in negotiating with international petroleum companies the governments of these countries lack the bargaining power that OPEC governments have, with their substantial known and probable reserves. Nevertheless, some of the governments of the OIDCs have sought to establish contract terms similar to those concluded by the large oil exporters, with the result that little foreign company exploration has been attracted.

The Ideological Factor in Contractual Arrangements

Our principal interest in this study is with those developing countries that require risk capital and technology for petroleum exploration and development, and with the types of contractual arrangements that have evolved. A basic problem for many of these countries has been strong political antagonism toward international petroleum companies and the desire of Third World governments to negotiate arrangements with foreign companies that will be in harmony with the United Nations' principle of "full permanent sovereignty of every State over its natural resources."[15] In some countries this ideological position has at times taken the extreme form of not allowing any foreign investment in petroleum production. This was true, for example, in the case of Brazil for several years, but the sharp rise in petroleum prices during the 1970s, together with the inability of the national petroleum company, Petrobras, to expand production, forced that country to change its policy. Argentina's petroleum policy has for decades shifted with the political philosophy of the government in power.[16] The popular antipathy toward international petroleum companies has been sufficiently strong to constrain both dictatorships and democracies in negotiating contracts with these companies. In some countries, the new forms of contractual arrangements have been designed to give the appearance of full control by the state over the country's petroleum resources and to avoid any resemblance to the traditional concession agreements.

The UN General Assembly has sought to support its resolutions relating to the full and permanent sovereignty of countries over their natural resources by proposing the establishment of international agencies that would provide both financing and technical assistance to developing countries that desire to explore and develop their petroleum resources *without equity investment by international petroleum companies*. However, as is discussed in chapter 12, the United States and other industrial countries have rejected these proposals. Although the United States and the World Bank are providing a certain amount of technical and financial assistance to

[15]This theme has been put forward in a series of UN General Assembly resolutions, including 523 (VI) of January 12, 1952; 1803 (XVII) of December 14, 1963; and 3201 (S-VI) and 3202 (S-VI) of May 1, 1974 on "The Declaration and Programme of Actions on the Establishment of a New International Economic Order (New York, United Nations).

[16]See Edwards, "The Frondizi Contracts," in Mikesell, *Foreign Investment*, chapter 7.

state petroleum enterprises for exploration, much of this assistance is designed to promote programs in the OIDCs to attract investment by international petroleum companies.

Current Contractual Arrangements Between Petroleum Companies and Host Governments

The principal categories of contractual arrangements currently employed between petroleum companies and host governments are (1) the concession agreement, (2) the joint venture agreement, (3) the production-sharing agreement, and (4) the service contract. Actually, there is no standardized format for any of these categories and each may contain some of the characteristics of the others. Moreover, several subcategories of each of the principal categories are recognized. The following discussion concentrates mainly on those arrangements that are characteristic of contracts between petroleum companies and host governments of countries represented by the case studies in part II which, with the exception of Indonesia, cover only the non-OPEC LDCs. However, a few examples are drawn from the North Sea contracts and those with OPEC countries in the Middle East and Africa.

The Concession Agreement

Although the *traditional* concession agreement is no longer employed, there are several non-OPEC LDCs, including Sudan, Thailand, and Tunisia, that employ the basic concession agreement format. Concession contracts or leases are also employed by the U.S. government for continental shelf exploration and production, and by the U.K. and Norway in the North Sea. The distinguishing features of concession agreements are the absence of direct government participation in operations and marketing, and the use of royalties, income taxes, and bonuses for payments to the government, as contrasted with some form of output sharing. The concession agreement grants the petroleum company the

right to explore, produce, and sell (including export) petroleum and/or natural gas within the concession area for a fixed period of time. There is usually a drilling obligation and/or an obligation to spend a certain sum of money each year and to relinquish a certain percentage of the concession area on a fixed time schedule. There is frequently a signature bonus, a discovery bonus, and a bonus on reaching a certain level of production, plus a royalty on output and an income tax which is often graduated. In some cases one or more of the bonuses may be recouped as advances on royalties or other fiscal payments.

Methods of granting concession contracts differ from country to country. Governments usually announce certain areas available for concessions and if any exploratory work has been done in the area by the government, interested companies may purchase copies of reports from the government. Many, if not all, the conditions of concession contracts may be established by the country's petroleum legislation, so that the scope for negotiation of the contracts will differ from country to country. Frequently governments prepare model contracts on the basis of existing legislation and award concessions on specific tracts by negotiation or by competitive bidding. Bids may be made on the basis of the size of the bonus, the level of exploration expenditures, the number of wells that will be drilled during the first few years of the contract, or, in a few cases, on the amount of royalty per barrel.

The concession agreement may stipulate that the concessionaire supply a certain amount of output to the domestic market at a price that may differ from world prices. Since government revenue is likely to be mainly a function of the contractor's revenue from oil exports, the host government would not ordinarily be interested in limiting the export price received by the concessionaire. Most concession agreements provide for revenues to be calculated at the world price for purposes of taxation. However, under some concession agreements, the government has established "posted" prices to be used in the calculation of income taxes that are often well above the world market price for petroleum. For example, Nigeria began setting posted prices for petroleum above the world market price during the world oil surplus that started in 1980, but early in 1982 reduced posted prices as a means of lowering per barrel taxes.[17]

The modern concession agreement usually provides for granting an exploration permit for a fixed period, and is renewable as long as the work requirements are met. A concession for exploitation is granted automatically upon a discovery, the conditions for which are defined in the agreement. The concession is usually for a substantial period of time, fifty to sixty years. Once a concession has been granted, the title holder has an obligation to exploit the concession to the maximum extent consistent with achieving the optimal yield and subject to reasonable conditions of profitability or "state of the art" practice. Special provisions are made in the event of a discovery of gas deposits.

Holders of exploration permits and concessions are usually required to submit work programs for exploration and exploitation prior to each year of activity. In addition, the government requires detailed reports, including geological data, on each year's activity.

The Joint Venture

Although there are a variety of possible provisions that enable governments to participate in managerial decisions under a concession agreement, such as approval of work programs and budgets, governments have tended to favor agreements that give them formal as well as *de facto* participation in management. The joint venture is the earliest arrangement employed to give governments formal participation with private companies in petroleum operations. According to Kamal Hossain, the earliest joint ventures were those entered into in 1957 by the Italian state oil corporation, ENI, and Egyptian and Iranian government oil enterprises (GOEs).[18]

As in the case of a joint venture involving two or more private companies, the distinguishing feature of the joint venture between a private company and a GOE is shared equity contributions to the enterprise. The equity may represent shares in an operating company, as was the case with the early ENI joint ventures mentioned above, or it may take the form of a partnership along the lines of the Contract of Association between the Colombian GOE, Ecopetrol, and private petroleum companies (see chapter 10). The joint venture may be initiated when exploration rights are granted, or it may result from a transfer of a portion of the equity of a foreign company that may have been established in the country for years. The latter case is often referred to as a "partial nationalization" and usually takes place at the "request" of the host government. In some cases, the assets of a privately owned corporation may be partially nationalized by the government rather than the government's taking equity shares in the company itself. Beginning in 1972, the Saudi Arabian GOE, Petromin, acquired 20 percent of the assets of Aramco, a company owned by four U.S. petroleum companies; this was later increased to 60 percent and recently to 100 percent.[19]

The joint venture involves joint ownership of assets and concession rights, a sharing of certain costs of operation, and a sharing (usually in proportion to the equity interests) of the net revenues. In cases where the joint-venture contract is established when the private company enters into exploration on a petroleum concession, there are in general two types of arrangements. The first is one in which the GOE does not share in the costs until a commercial field has been discovered, so that joint cost-sharing does not occur until after the development and exploitation periods. The private company bears the risks and costs of exploration and is not compensated for any part of the exploration costs incurred prior to the time that the existence of a com-

---

[17]Nigeria's older contracts are regarded as concession contracts, but beginning in 1979 the new contracts have some of the characteristics of both production-sharing and service contracts.

[18]Kamal Hossain, *Law and Policy in Petroleum Development* (New York, Nichols Publishing, 1979), p. 121.

[19]In 1980 the Saudi Arabian government took over the remaining operating assets of Aramco, which ceased to be a producer, but operates Saudi facilities for a fee. Aramco's owners are Exxon, Mobil, Socal, and Texaco.

mercial discovery is mutually accepted. This type of arrangement is represented by the Contracts of Association between Ecopetrol and private petroleum companies operating in Colombia.

Under the second type, the GOE shares in costs and risks during exploration as well as during the development and exploitation periods. This type of arrangement is represented by the joint-venture, production-sharing contracts negotiated by the Indonesian GOE, Pertamina, and private petroleum companies (see chapter 5). Even when the GOE shares in development costs, the arrangement may provide that the GOE's share is paid out of its share of future revenue or output so that it is not required to put up funds in advance.

Under the arrangement whereby the GOE shares in the risks of exploration, the government can generally obtain a larger share of net revenues from the operation than where it shares only in development and production costs after discovery has been made.

In the joint venture, the private company is almost always designated as the operator, but the GOE usually participates in management through a joint-management committee which approves the budget and work programs submitted periodically by the operator and deals with other managerial matters. The private company is entitled to its share of the output, but is subject to income taxes on that output. The private company may also market the GOE's share of the output at the option of the GOE.

Production-Sharing Contract

The production-sharing contract (PSC) was first employed in 1966 by Indonesia in the form of a contract between that country's GOE and a foreign-private petroleum company. The PSC provides for production sharing, but without equity and cost sharing, as in the case of the joint venture. In its initial form the PSC avoided the problems associated with levying income taxes by having them paid by the GOE. Where a fixed amount of oil was allowed to cover production costs to be deducted from total output before production sharing, or where the share going to the contractor included compensation for all costs,

it was not necessary for the GOE to determine either the contractor's costs or the appropriate price for the output.[20] However, this relatively simple arrangement has been changed substantially in recent years so that the PSC has lost much of its original character.

Since its initial use by Indonesia, various forms of production-sharing contracts have been employed by Chile, Guatemala, Israel, Ivory Coast, Egypt, India, Peru, Libya, Malaysia, Syria, Trinidad, Oman, Sudan, and several other LDCs. In the early simplified models, the PSC gave both the host government and the private company a share of the output that each could market as it pleased, thereby avoiding all the complexities of calculating costs and net revenues for determining tax obligations. However, when petroleum prices increased in 1973, governments wanted contracts that enabled them to increase their share of net profits as prices rose. They therefore demanded a renegotiation of the old PSCs and revised their model contracts for the negotiation of new ones. Another major factor leading to the change was the ruling by the U.S. Treasury in 1977 that in order for foreign taxes to be credited against the tax obligations of U.S. corporations, the foreign taxes had to be paid directly to foreign governments on net revenues, and a clear distinction made between taxes on net revenues and payments to governments for the value of their resources produced and sold (see chapter 11). Although a number of so-called PSCs have been negotiated in recent years, it can be said that their basic character has been changed to the point where they represent a combination of revenue-sharing arrangements in which production sharing is only one element. This is well illustrated by the evolution of Indonesian PSCs described in chapter 5.

The Service Contract

The term "service contract" has been applied to so many different types of agreements that em-

[20]Peru's first production-sharing contract excluded an allowance for cost recovery and provided that the GOE, Petroperu, absorb the contractor's income tax.

body elements of other more definitive classifications that the term is more semantic than substantive. The *pure* service contract is one under which a private petroleum company agrees to perform certain specified services for the government or a GOE in return for fixed payments. For example, some of the Argentine service contracts with oil drilling firms provided for a fixed payment for drilling a given number of wells at a specified depth, whether or not any oil was produced.

Because of widespread public opposition to the traditional petroleum concession, some GOEs established in developing countries were given a monopoly on all petroleum-producing operations, but because of their limited technical capacity, they frequently hired the services of private companies with appropriate skills and equipment. However, the GOEs were often financially unable to assume the risks of expensive exploration or were unable to obtain the services of international petroleum companies with the most advanced technology for exploration on a fixed fee basis. Governments therefore devised contracts under which the risks and costs of exploration were undertaken by the contractor in return for a certain amount of petroleum, or

a portion of the revenue from the sale of petroleum. These contracts were called service contracts or, in some cases, risk-service contracts. Since the contractor receives nothing if oil is not discovered and produced, such so-called service contracts are in many ways similar to a concession agreement with substantial involvement in production and marketing by the government owned enterprise.

Examples of service contracts are provided in the case studies on Argentine and Brazilian contracts in part II. Some of the Argentine contracts with private petroleum companies provided for operations in areas where reserves had been proved and the companies were compensated by a certain percentage of the oil produced at prices fixed by the government. Such contracts involved relatively little risk to the contractors. Later Argentine risk-service contracts, as well as Brazilian risk contracts, called for exploration in areas where reserves had not been proved. In such cases the service contracts bore considerable resemblance to concession contracts or PSCs. The contractor makes a risky investment and his revenue after taxes depends upon an agreed formula for the sharing of revenues, provided the outcome of the investment is successful.

4

The Effects of Alternative Fiscal Arrangements on Host Country Returns and Petroleum Investments

Introduction

A host government may have several objectives in negotiating contracts for exploration and development of potential oil lands, but we concentrate on the following: (1) maximizing the amount of exploration to increase the chances of commercial discovery and to obtain technical information on contract areas; (2) maximizing the efficiency of the development of oil reserves; and (3) maximizing the host government's revenue from oil lands. These three objectives are interrelated in considerable measure. The first two are important for maximizing total revenues over costs from a large area that may contain several oil reservoirs, while the third objective has to do with the division of the economic rent between the host government and the petroleum company or companies. By economic rent we mean the surplus of revenue over full economic costs of producing oil.

The first part of this chapter provides an analysis of (1) the criteria for petroleum investment decisions under conditions of risk and uncertainty of returns; (2) the effects on government revenues of alternative methods of taxation; and (3) the effects of the tax system on the devel-

30

opment of oil fields of different sizes. It should be emphasized that in (1) the analysis is designed to summarize the principal factors in investment decisions that have a bearing on the effects of alternative contractual arrangements. It does not purport to provide a systematic review of the entire investment decision-making process in petroleum.

This general analytical discussion is then applied to competitive bidding systems; nonbidding contracts employing different types of tax regimes, bonuses, and revenue-sharing arrangements; and various types of government equity participation arrangements.

General Analysis

The Criteria for Investment Decisions

PROFIT CRITERIA. Most investors employ the internal rate of return (IRR) as the basis for making an investment decision or of choosing among alternative investment opportunities. The IRR is the discount rate that equates the present value of the stream of revenues from a project with the present value of the costs, including taxes. The basic formula for defining the IRR is:

$$\Sigma_t R_t (1 + r)^{-1} = \Sigma_t C_t (1 + r)^{-1}$$

where r is the IRR, R_t is the positive cash flow (revenue) during the life of the project, t, and C_t is the negative cash flow (cost) during the period t. In a petroleum investment, large expenditures for exploration and development are made over a period of time before there is any revenue. In some cases these outlays include initial bonus payments to the government. After revenues begin to flow, operating costs and various forms of taxes are incurred; these usually are a function of revenues or of revenues less costs, as in the case of taxes on profits. Therefore, in the formula above, R_t may be net cash flow or current revenues less current costs, while C_t may be the initial capital costs of the project.

A variation of the IRR profit criteria is net present value (NPV), which is the difference between the present value of the positive cash flow over the period, t, and the negative cash flow for a *given* rate of discount, d. The formula for NPV is:

$$\Sigma_t R_t (1 + d)^{-1} = \Sigma_t C_t (1 + d)^{-1}$$

NPV may be negative or positive, but a negative NPV implies that the internal rate of return on the project is less than d, which may be the investor's minimum acceptable rate of return or IRR. If NPV is equal to zero, d is just equal to the IRR on the project. Both concepts are useful in analyzing the criteria for investment decisions, but they are based on the same principles.

The calculation of an *expected* IRR or NPV begins with a cash flow analysis of projected revenues and costs for the project. During the initial years of a petroleum investment, there will be outlays for exploration, including seismic surveys and one or more exploration wells. If the exploration is successful in terms of the reserves delineated, investment will be made in producing wells, platforms, pipelines, and various types of infrastructure. The required investments for the development stage will depend on the size of the field discovered in terms of the volume of reserves

unless the investment is being made in a field in which reserves have already been proven. Revenues (output times price) must be projected for fields of varying size and development, and operating costs (which differ with field size) must also be projected. Various taxes must be projected in the cash flow analysis, including royalties, profits taxes, etc., in accordance with the government's tax regime. Profits tax regulations usually provide for the rate at which capital investment, including exploration outlays, can be depreciated, since depreciation is ordinarily an allowable deduction from gross revenues (revenues less operating costs) in calculating taxable profits. It will then be possible to calculate net cash flow (NCF) for each year during the life of the project, and, given the NFCs, NPV can be calculated for a given rate of discount, d, which usually reflects the opportunity cost of the investor's capital. In addition, the IRR can be calculated from the NFCs.

This highly simplified explanation is illustrated in the hypothetical example in table 4-1, which assumes no royalty, a 50 percent tax on taxable income, and a 15 percent rate of discount for the NPV. It will be noted that since NPV is positive, the IRR exceeds the investor's discount rate of 15 percent. However, this base case projection takes no account of the probability that a field of this size will be discovered. A larger field will yield a larger revenue and larger operating costs, and development outlays will also be larger because of the need for more producing wells and other equipment. However, investment costs will not ordinarily rise in proportion to the increase in revenues, so that NPV (and IRR) will be higher the larger the field discovered. As will be discussed below, the probability of discovery tends to decline with the size of the field. It should be emphasized that the hypothetical example given in table 4-1 represents a base case projection for a field of a particular size without taking into account the probability of discovery.

THE NPV OF A PROJECT TO THE HOST GOVERNMENT. NPV may also be employed to measure the value of the same project to the host government under alternative fiscal arrangements with the investor. Assuming the government has no equity or loan investment in

Table 4-1. Hypothetical Example of Calculation of Net Present Value (NPV) and Internal Rate of Return (IRR)

(millions of dollars; ( ) = negative cash flow)

	1	2	3	4	5	6	7	8	9	10	11	12	13	14	15	16	17	18	19	20
(1) Exploration outlays	(10)	(10)	(10)																	
(2) Development investment				(35)	(35)															
(3) Revenue						100	100	100	100	100	100	100	100	100	100	100	100	100	100	100
(4) Operating costs						(20)	(20)	(20)	(20)	(20)	(20)	(20)	(20)	(20)	(20)	(20)	(20)	(20)	(20)	(20)
(5) Allowable depreciation (20% per year)						20	20	20	20	20										
(6) Taxable income						60	60	60	60	60	80	80	80	80	80	80	80	80	80	80
(7) Tax (50% of taxable income)						(30)	(30)	(30)	(30)	(30)	(40)	(40)	(40)	(40)	(40)	(40)	(40)	(40)	(40)	(40)
(8) Net cash flow (3)-(4)-(7)	(10)	(10)	(10)	(35)	(35)	50	50	50	50	50	40	40	40	40	40	40	40	40	40	40

NPV at 15 percent = $72 million

IRR ≅ 31 percent

the project, the NPV of the project to the government for a particular fiscal arrangement is the present value of the expected stream of government revenues, discounted at the government's rate of discount or social rate of discount. The same aggregate revenue accruing to the government over the life of the project could yield different NPVs, depending upon the time pattern of the revenues under different fiscal regimes, e.g., a signature bonus versus royalties. Various sources of government revenue may also be adjusted for risk or the probability that the revenues will be realized. For example, bonus payments made at the time the contract with the investor is signed involve no risk, while royalties depend upon the discovery of oil and the size of the fields discovered. As will be noted below, different fiscal regimes involve a different sharing of the risk between the government and the investor.

UNCERTAINTY AND RISK. By its very nature, petroleum exploration involves risk since the outcome cannot be known with certainty. However, we may distinguish between two types of uncertainty situations. The first type exists when the probability distribution of the possible outcomes is known with reasonable certainty. In calculating the probability distribution of possible outcomes from a petroleum investment, it is necessary to take into account a large number of variables, each of which has its own probability distribution. An

important variable has to do with the estimated volume and distribution of reserves in the area of exploration. Knowledge of the geologic structure and experience with exploring similar structures will provide a basis for determining the probability of discovery of *any* petroleum reserves, while previous exploration and production in a given region provides the basis for estimating the size/frequency distribution of reserves in a given field, i.e., the probability of the existence of reserves of differing amounts. Normally the probability of discovery is smallest for the largest fields so that the probability of finding a field capable of producing over 100,000 bpd may be only 5 percent, while the probability of finding a field capable of producing 5,000 bpd may be 25 percent, and the probability of finding only dry holes is 75 percent.[1]

Given the other variables in a base case cash flow analysis for fields of different sizes and,

[1] For a good discussion of this subject, see Charles G. Johnson, "Establishing an Effective Production-Sharing Type Regime for Petroleum," *Resources Policy* (June 1981) pp. 131–132; see also *International Petroleum Encyclopedia* (Tulsa, Okla., Petroleum Publishing Co., 1979) vol. 12, pp. 196–219; C. Jackson Grayson, Jr., *Decisions Under Uncertainty: Drilling Decisions by Oil and Gas Operators* (Boston, Mass., Harvard Business School, 1960); and R. W. Megill, *An Introduction to Exploration Economics* (Tulsa, Okla., Petroleum Publishing Co., 1971). For data on discovered field sizes in various countries from which field size distributions for regions can be calculated, see *International Petroleum Encyclopedia* (Tulsa, Okla., PennWell Publishing Co., 1982) vol. 13, pp. 229–249.

hence, different potential revenues, a probability-weighted NPV can be calculated by applying the appropriate probability coefficient to the NPV for each of several field sizes. If the NPV for a field capable of producing 100,000 bpd is $500 million and there is a 5 percent probability of discovery, the probability-weighted NPV is $25 million. If the NPV for a field producing 5,000 bpd is $16 million and has a probability of discovery of 25 percent, the probability-weighted NPV is $4 million. The probability-weighted NPV for the investment must be determined by combining the probability-weighted NPVs for a number of possible discoveries.

Although the highest degree of uncertainty is in the geologic outcome, other variables important in the base case cash flow analysis are also subject to uncertainty. There is the possibility of investment cost overruns or that operating costs may be understated. Future petroleum prices are also uncertain and they should be estimated in terms of a range of prices, each with its own probability coefficient. There is also political risk, e.g., the terms of the contract with the host government may be changed by forced renegotiation or the government may expropriate the producing field. Elaborate models have been formulated for dealing with a large number of variables, each with its own probability distribution, for calculating the probability coefficient for each of several final outcomes in terms of NPV and the overall NPV for the investment.[2] On the basis of this information, the investor can determine whether the overall probability-weighted NPV or IRR is high enough to warrant an investment. The probability-weighted NPV or IRR is frequently referred to as the *expected NPV* or *expected IRR* on the investment and this designation will be employed in the discussion that follows.

The range of possible outcomes from an exploration program may run from nothing but dry

holes to a bonanza. Knowledge of the probability coefficient for each of a number of possible outcomes would enable the investor to estimate the probability-weighted or *expected* IRR or NPV on the investment. (The *expected* IRR is the average of a number of possible internal rates of return, each weighted by the probability that the particular rate of return will be achieved.) However, it is not always possible to estimate the probability coefficients for each possible outcome except on a broad basis. For example, we might assume a "normal" distribution in which each outcome has an equal probability within a range of, say, 10 to 30 percent. On the other hand, the geologists familiar with the geological structure of the region might expect a triangular distribution, i.e., one in which there is a greater chance of each outcome within a given range being above rather than below the mode (or vice versa). In this case, coefficients will be assigned accordingly and the *expected* IRR calculated.

The second type of risk situation exists when the probability distribution of possible outcomes cannot be estimated except within wide limits of confidence, or not at all. This situation would exist for an initial investment in seismic activity and exploratory drilling in an area where very little is known. Even after an area has been explored and one or two successful wells drilled, the probability distribution of the productivity of additional wells will be uncertain pending the results of further drilling and production. There is also uncertainty regarding other variables, such as petroleum prices, that go into calculations of the probability distribution of the financial outcome of an investment in terms of NPV or IRR. The degree of certainty regarding the distribution of possible financial outcomes is an important factor in determining the discount rate a company will use in calculating the NPV of the investment. When a company has no basis for estimating the probability of possible financial outcomes, the prospective investment becomes a pure gamble. However, petroleum companies do not make large investments unless they know something about the geology of the area. A large company might spend a few hundred thousand dollars on a seismic survey in order to

[2]See J. W. Whitney and R. D. Whitney, *Investment and Risk Analysis in the Minerals Industry* (Reno, Nev., John W. Whitney, Inc., 1978) chapter 5. See also, G. M. Kaufman, *Statistical Decisions and Related Techniques in Oil and Gas Exploration* (Englewood Cliffs, N.J., Prentice-Hall, 1963); and F. J. Stermole, *Economic Evaluation and Investment Decision Methods* (Golden, Col., Investment Evaluations Corp., 1980).

acquire information. If the results appear favorable, a large company or a consortium might invest a few million dollars for exploration without having a reasonable basis for estimating whether the odds of finding a commercial deposit are 10 to 1 or 20 to 1. If a company were making a $100 million investment under conditions of substantial uncertainty regarding the probability distribution of the outcomes, it would be likely to employ a very high discount rate for calculating the net present value of the investment opportunity.

Let us assume that the *expected* IRR on a prospective petroleum investment is satisfactory to the investor in terms of the opportunity cost of his capital, or minimum acceptable rate of return. Unless an investor is *risk neutral*,[3] he will not make the investment unless the expected rate of return is sufficiently above his minimum acceptable rate of return to compensate for his risk aversion. For example, a risk-averse investor would not make an investment that has a 50 percent chance of being 25 percentage points higher or lower than his acceptable rate of return. But he might make the investment if the higher rate of return has a 75 percent chance of being realized as against only 25 percent for the less than acceptable rate of return. In other words, the risk-averse investor will demand odds in his favor. However, potential investors have different degrees of risk aversion, assign different probability coefficients to possible financial outcomes, and have different acceptable rates of return.

The odds required by a risk-averse investor may be regarded as a form of risk premium. The risk premium may vary with the wealth of the investor even for the identical *expected* rate of return. The risk premium required by an investor who is contemplating making an investment that represents a large proportion of his total wealth will normally be higher than that for, say, a multibillion dollar corporation contemplating a risky investment of $25 mil-

lion.[4] A large company will normally undertake a number of explorations in different regions of the world to maximize its average return, expecting there will be a few investments with very large returns to offset the many with no positive returns or less than average returns. This has been called the *portfolio effect*. For example, a large firm may undertake a dozen or more exploration ventures where there is no more than a 15 percent chance of a discovery, but a 10 percent chance that the returns will be 10 times the acceptable return. A small firm contemplating only one or two investments a year, each of which constitutes a substantial portion of its total assets, will require not only better odds (and therefore require a higher expected IRR), but will be attracted to investments with a high probability of outcomes approximating its acceptable rate of return, rather than gambling for very high stakes. These considerations tend to give large firms a competitive advantage in bidding for petroleum leases with low success ratios but with very high returns on the winners. Smaller companies can, of course, overcome this disadvantage by forming a consortium in which each company would have no more than, say, 25 percent of the equity, but this would sacrifice independent management of the operation.

When account is taken of risk aversion, it is possible to calculate a *risk-corrected (expected) internal rate of return*. The risk-corrected internal rate of return is a function of the investor's coefficient of risk aversion and the *variance* of the rate of return. This measure, which can be used for comparing alternative tax systems, is discussed in a subsequent section of this chapter and the method of calculation is given in appendix A.

---

[3]Risk neutrality means willingness to accept the odds represented by the probability-weighted outcome, e.g., making an even bet based on flipping a coin.

[4]For a discussion of risk premium on petroleum investments, see Steven W. Millsaps and Mack Ott, "Risk Aversion, Risk Sharing, and Consortium Bidding: A Study of Outer Continental Shelf Auctions," (mimeo) December 1980. (Millsaps is at the Appalachian State University, Boone, North Carolina, and Ott is at Pennsylvania State University, University Park, Pennsylvania.) See also, Kenneth J. Arrow, *Essays in the Theory of Risk Bearing* (Amsterdam, Holland, North Holland Publishing, 1970).

Risk and Profits Criteria for Different Stages of a Petroleum Investment

Although the exploration period entails the greatest risk or uncertainty in a petroleum investment, the required outlays at this stage are lower than in the development stage. If exploration does not result in the discovery of, reserves, no further outlays are made. Some petroleum contracts with host governments require that a certain number of exploratory wells be drilled (or equivalent payments be made to the host government) even though the seismic surveys, and perhaps an initial exploratory well, do not produce results that warrant further exploration. In other contracts the investor can walk away from the project after a minimum amount of exploration indicates a very low probability of discovery. This does not mean, however, that other contractual conditions relating to the division of revenues between the investor and host government are irrelevant for the decision to undertake exploration. First, the decision to invest in exploration depends upon the expected NPV or IRR in the event any commercial reserves are discovered. Second, the expected NPV or IRR will differ substantially for any given set of contractual conditions with the amount of reserves discovered. The contractual conditions will determine whether it is profitable to produce a field below a certain size, and the probability of finding smaller reserves is almost always greater than that for finding larger reserves. Therefore, if contractual conditions are such that an acceptable IRR could be realized only in the (less probable) event that a very large volume of reserves were discovered, the exploration may not be undertaken.

The analysis of profit criteria and risk in this section has important implications for the realization of host country objectives and for the evaluation of alternative types of contracts between the host government and petroleum companies from the standpoint of the maximization of host country objectives. The provisions of contracts affect the estimates of variables in the calculation of expected NPVs of government revenues. Certain contractual forms may be more advantageous to governments than others, de-

pending upon the timing and degree of certainty of different sources of revenue.

The Effects on Government Revenues of Three Basic Fiscal Mechanisms

The several categories of contracts between petroleum companies and governments described in subsequent chapters of this book employ a mixture of the basic taxing or revenue-sharing mechanisms, each of which has certain effects on investment decisions and methods of operations by the petroleum producers, and on the maximum economic rent from a project that the government will be able to extract.[5] Although variations in these basic fiscal mechanisms will be examined, the following paragraphs illustrate the effects on government revenues of three basic fiscal arrangements for essentially the same project: (1) a signature bonus payable at the time of the negotiation of the contract; (2) a tax on net profits; and (3) royalties or any payments as a proportion of gross revenues, including production sharing. The difference in the amount of government revenue in each case arises basically from differences in the sharing of risk between the contractor and the government.

BONUS. Let us assume that the expected net present value of a successful commercial oil field is estimated by a prospective contractor at $400 million, including allowance for exploration, development, and operating costs, with a 25 percent probability of success, and that exploration costs for the field (successful or not) are $64 million. We shall also assume that the discount rate for calculating the present value is the opportunity cost of capital to the contractor in other employments, and that the contractor is risk neutral. The expected NPV of the project, excluding any bonus, royalty, or profit sharing, may be expressed as follows:

$$NPV = \frac{1}{4} (\$400 \text{ million}) - \frac{3}{4} (\$64 \text{ million}) = \$52 \text{ million}.$$

Fifty-two million dollars is the maximum rent

[5]The economic rent is the difference between total revenues and total economic costs, including the minimum acceptable return on the investment.

that could be extracted by the government with a 25 percent probability of success.[6]

Let us assume the government imposes a signature bonus of $50 million and no other charges. The expected NPV of the project then is only $2 million, but if the contractor is risk averse, a $50 million bonus may not leave him enough to undertake the project. In this case, the contractor bears the entire risk of the project, including payment of the bonus whether he is successful or not.

NET PROFITS TAX.   Let us now assume that the government imposes a 20 percent tax on net profits. This would be roughly equal to $80 million, or 20 percent of the expected NPV of the field. In this case the expected NPV is:

$$\frac{1}{4} (\$400 \text{ million} - \$80 \text{ million}) - \frac{3}{4} \times \$64 \text{ million} = \$32 \text{ million}$$

By sharing the risk with the contractor and assuming the project is successful, the government obtains $80 million and provides the contractor with an expected value for the project of $32 million.

ROYALTIES.   Now let us assume that the government imposes a royalty of 15 percent on the gross revenues instead of a bonus or profit sharing. Let us further assume that the present value of the oil produced by a successful field is estimated at $540 million and the present value of the cost is $140 million, so that the net present value of the project is again $400 million. A 15 percent royalty on gross value of output is roughly equal to $80 million, provided the contractor produced the same amount of oil, or $540 million worth. With the 15 percent royalty, the expected NPV of the project to the contractor would again be about $32 million, while the government obtains $80 million. However, the royalty might lead the contractor to reduce output by, say, 5 percent, assuming it would not pay to produce the marginal output at a royalty of 15 percent. This would result in a reduction of government revenue to $77 mil-

lion, while perhaps reducing the expected value of the project to the contractor as well. The same analysis would apply to a production-sharing agreement instead of a royalty.

The analysis of the three examples given above implicitly assumes zero time preference for the government, i.e., the government is indifferent between early revenue and revenue later on. However, the analysis could be adjusted to take into account an appropriate rate of time preference for the government. This analysis also implicitly assumes that the government as well as the contractor is risk neutral.

Effects of the Tax System on the Development of Oil Fields of Different Sizes

In the typical situation where a petroleum firm explores an area under a license or a concession, there is the possibility of finding fields of different sizes in terms of productive reserves (cut-off field size), with the probability of discovery declining with the increase in field size.[7] Let us assume an area with two possibilities—a small field with a 20 percent probability of occurrence, and a larger field of twice the cut-off field size with a 10 percent probability of occurrence. However, the pretax expected NPV of the smaller field may be less than half that of the larger field because of lower unit costs for the larger field resulting from scale economies.

A fixed royalty per unit of output might result in the present value of the tax burden on the smaller field being greater than the pretax expected NPV of that field, while the present value of the tax burden on the larger field might be significantly less than the pretax expected NPV of that field. In this case the petroleum firm's expected NPV of the investment would be based solely on the discovery and production of the larger field. However, if the tax burden were distributed between the two fields so that both

---

[6]The net present value of a successful outcome allows for the cost of exploration. Hence, one-quarter of the exploration cost is already accounted for in the calculation of he expected present value of the project.

[7]The material in this section is based in part on a paper by Thomas R. Stauffer and John Gault, "Effects of Petroleum Tax Design upon Exploration and Development" (presented at the 1981 Economics and Evaluation Symposium of the Society of Petroleum Engineers of the AIME, Dallas, Tex., February 25–27, 1981).

fields would be produced, the combined expected NPV would be higher than in the first case; the revenues accruing to the government could be larger; and the petroleum firm's risk perception would be reduced.

The situation described above may be illustrated by the following example.[8] Assume the pretax present value of the smaller field is 600 and that of the larger field is 2,000. Also assume a tax system (tax system A) in which the present value of the tax burden is 600 for the smaller field and 1,000 for the larger field. In this case, the after-tax expected NPV (or present value of the probability-adjusted cash flow) for the smaller field is zero, and the expected NPV for the investment is 100 (assuming a probability coefficient of 0.1). Assuming only the larger field would be produced, the expected NPV of the government's revenue is also 100.

Now let us assume a tax system (tax system B) under which the present value of the tax burden for the smaller field is reduced to 400 and the present value of the tax burden on the larger field is 1,400. In this case the present value of the after-tax cash flow for the smaller field is 200 and the expected NPV for that field is 40 (assuming a probability coefficient of 0.2). For the larger field the present value of the after-tax cash flow is 600 and the expected after-tax NPV is 60. The combined expected NPV for the two fields is 100 under tax system B, the same as under tax system A. However, under tax system A, the expected present value of the government's tax revenue is 100, while under tax system B the expected present value of the government's tax revenue is 220 (80 from the smaller field and 140 from the larger field).

Under tax system A, the coefficient of variance (standard deviation) is increased relative to that for tax system B, which permits both fields to be produced. Since the *variance* of the present value of the probability-adjusted cash flow (expected NPV) under tax system B is about one-third that for tax system A, the risk perception to the petroleum firm is substantially less under tax system B and, therefore, the *risk-*

corrected IRR is potentially greater under tax system B. If we assume that the probability-adjusted (expected) IRR is 40 percent for both fields, and that the investor's coefficient of risk aversion (which measures his attitude toward risk) is 0.35, the *risk-corrected* IRR for tax system B is 28.1 percent, and for tax system A is 16.9 percent.[9] Therefore, under tax system A the contractor might not undertake the investment if his minimum acceptable IRR (before adjustment for risk aversion) were 20 percent.

The important point to be noted from the above analysis is that any tax system that places a sufficiently high burden on smaller or marginal fields so as to preclude their development will prove to be a less efficient revenue producer for the government. In addition, it will reduce the risk-corrected internal rate of return of the investor, compared with a tax system that permits smaller or marginal fields to be produced.

Alternative tax systems yield differing results from the standpoint of both risk and government revenue generation for the same expected IRR for the producer. A bonus system places a heavy tax burden on small or marginal fields and, therefore, increases the variance of the rate of return relative to most other tax systems. Thus a bonus system reduces the risk-corrected IRR from the expected IRR by a higher percentage than does, say, an income tax or royalty. If the bonus is set by competitive bidding, the difference between the expected IRR and the desired risk-corrected IRR will be reflected in a reduction in the size of the bid by risk-averse bidders. A high excess profits tax eliminates the possibility of a high return on low-probability large fields with high pretax present values. By increasing the variance of the IRR (relative to that for a uniform income tax rate), an excess profits tax reduces the risk-corrected IRR below the expected IRR by a substantially greater percentage than would be the case for a uniform income tax rate.

In the case of a uniform income tax rate or a uniform royalty on gross revenue, the former is likely to place a relatively smaller tax burden on a small field with higher unit costs than on

---

[8]This example is based on one given by Stauffer and Gault in "Effects of Petroleum Tax Design."

[9]See appendix A for method of calculating variance and risk-corrected rate of return.

a larger field with lower costs as a consequence of scale economies. Hence, a uniform income tax rate is less likely to preclude the development of small or marginal fields than a uniform royalty rate. However, the possibility of eliminating production of small or marginal fields could be reduced by a sliding-scale royalty system that favored marginal fields.

It should also be emphasized that a tax system that maximizes production for the same expected rate of return to the producer will tend to maximize government revenue. However, in some cases reducing risk to the investor will increase the government's risk, but this may be offset by higher potential government revenue.

Application

Competitive Bidding

Some of the LDC petroleum-producing countries with which we are concerned in this book provide for competitive bidding on certain types of contracts. These include Argentina, Brazil, Guatemala, and Indonesia. However, the United States has had the greatest amount of experience with competitive bidding on oil and gas leases and this method of allocation is required for certain publicly owned onshore lands and all sales of leases on the outer continental shelf (OCS). Some Canadian provinces employ competitive bidding for oil and gas leases and an auction system has been used to a limited extent in the North Sea.[10]

There are several types of competitive bidding. Bonus bidding has predominated in the United States, and is also used in Brazil and Indonesia. Other types of bidding used in contracts include (1) the level of exploration expenditures (Argentina, Guatemala, and Indonesia); (2) the amount of oil or revenue retained by the contractor (Brazil and Guatemala); (3) the number of wells the contractor undertakes to drill (Brazil and Guatemala); and (4) the price paid by the government to the contractor for the oil produced (Argentina). There are still other types of competitive bidding that have not actually been used in contracts but have been discussed in the literature, e.g., the profit share bid. Some of these are discussed in the following sections.

BONUS BIDDING. In making a bonus bid, the petroleum company calculates the *expected* NPV of the lease or contract from the estimated recoverable oil in the tract on the basis of the expected future price of oil, the cost of extraction, the royalty rate and other taxes, and the probability coefficients of achieving a series of outcomes.[11] The amount of the bid reduces the *expected* NPV of the tract, but the maximum bid will normally not reduce the *expected* NPV below the amount that would yield an acceptable return to the bidder, taking into account his risk premium. (See the hypothetical case in appendix B.) Ideally from the standpoint of the host government, companies would bid an amount equal to the difference between the *expected* NPV of the lease and the risk-corrected NPV that reflects a normal rate of return on the company's aggregate investment in leases. However, such an outcome would depend on the number of competitors and the degree to which the bidders were risk averse.

As has been noted, the rates of discount employed will reflect the uncertainty of the probability distribution of possible outcomes as perceived by each bidder, even though all bidders may possess the same information on the particular tract on which bidding takes place. Each bidder will also take into account his perception of the maximum bid by the other bidders. Some bidders will be less risk averse than others, so that *risk-corrected* acceptable rates of return will differ.

Competition is increased by the number of

For a discussion of the British experience with auction systems in the North Sea, see Kenneth W. Dam, *Oil Resources* (Chicago, Ill., University of Chicago Press, 1976) chapter 4. Dam points out that the British government used the auction system only to a limited degree because it tended to favor U.S. over British firms in the allocation of concessions and there were strong political reasons for wanting to favor British firms.

[11]In most cases there will be several possible financial outcomes, each with a probability coefficient.

firms that may be bidding. The more firms that can be attracted to making bids, the higher the maximum bid is likely to be. D. K. Reece has shown that the higher the degree of certainty and the more companies that are engaged in bidding, the larger the economic rents that will accrue to the host government.[12] It should be cautioned, however, that Reece makes certain assumptions regarding the probability distribution of the estimated values of the lease and the bidding behavior of the firms that may not conform to reality. Nevertheless, there are intuitive reasons for the general validity of his conclusion.

Bonus bidding has the advantage of creating an environment for the most efficient exploration and development of the lease since the bonus is a sunk cost and therefore will have no effect on the subsequent operating strategy of the successful bidder. Bonus bidding is perhaps the most desirable technique where there is a substantial amount of information about the area that is to be leased and where there are a fair number of competitive bidders. For example, usually a substantial number of large companies bid for tracts in well-known areas such as the U.S. Gulf Coast. An analysis of 839 U.S. government offshore oil and gas tracts leased in the Gulf of Mexico between 1954 and 1962 showed that the average before-tax internal rate of return on these leases was only 9.5 percent. This compares with the average before-tax rate of return for all U.S. manufacturing firms over the 1954–76 period of 19.2 percent.[13] This suggests that some of the companies were overbidding in terms of a proper evaluation of expected returns. Bo-

nus bidding may also have an advantage where the government has a relatively high social rate of discount and/or is relatively risk averse.

Competitive bonus bidding has certain disadvantages, especially in the case of areas for which there is little geological information, and where the government is anxious to attract companies to undertake high-risk exploration in regions where no oil has been discovered. This has proved to be the case in certain Latin American countries where the governments have had great difficulty in attracting companies to bid on tracts put up for lease or contract.

The principal disadvantage of the bonus bid—or for that matter a fixed or negotiated bonus without bidding—is that it commits the company to a large fixed cost, and therefore greatly increases the cost of obtaining knowledge through exploration. A large bonus substantially reduces the *expected* NPV of the lease, in contrast to the same expected revenue from a royalty or profits tax that would be paid only after production was initiated. As has been noted, the discounted value of the cost incurred with a bonus payment is substantially higher than the discounted value of royalty or profits taxes paid after production is initiated. Moreover, in the absence of a bonus payment, the contractor might be willing to spend more on exploration, thereby buying knowledge available to both the contractor and the host government. A bonus payment designed to capture a portion of the rent requires the contractor to accept the entire risk of there being any economic rent to be derived from the lease for that portion of the investment represented by the bonus payment.

Another disadvantage of bonus bidding, particularly for areas where there is little knowledge, is that there may be little relationship between the actual economic rent from the project and the bonus. Not only is there no economic rent produced by dry holes, but the bonus may prove to be a small fraction of the economic rent produced by a bonanza. Of course, if a bonanza is found, the government may decide to change the terms of the lease. But such actions reduce the credibility of the host government and thereby increase the political risk factor which companies take into account when de-

[12]See Douglas K. Reece, "Competition Bidding for Offshore Petroleum Leases," *Bell Journal of Economics* (Autumn 1978) pp. 369–384. Reece's model assumes that the estimate of the gross value of the tract by each firm is drawn from a lognormal distribution of random variables. (If random variable Y has a normal distribution and $X = e^Y$, then X has a lognormal distribution, $Y = \ln(X)$, and e is the exponential constant = 2.71828.)

[13]See R. O. Jones, W. J. Mead, and P. E. Sorensen, "Economic Issues in Oil Shale Leasing Policy," James H. Gary, ed., *Eleventh Oil Shale Symposium Proceedings* (Golden, Col., Colorado School of Mines, November 1978), p. 205. The two rates of return many not be wholly comparable since the earnings of U.S. manufacturing firms were accounting rates of return rather than internal rates of return.

ciding whether to make an investment. When Occidental Petroleum discovered oil in the eastern jungle of Peru after many millions of dollars had been spent by a dozen other companies without a discovery, Occidental was rewarded by a drastic revision in its contract with the Peruvian government, which substantially reduced its potential earnings and, incidentally, its incentive to drill more wells.

A final disadvantage of bonus bidding is that smaller companies often cannot afford to bid, either because they cannot obtain the financial resources, or because they are not making enough risk investments in other areas of the world to operate on the basis of having a few very profitable ventures compensate for a number of failures. This will greatly reduce the number of bidders.

Governments are likely to obtain a higher proportion of the rents from bonus bidding if more information on the tract is available to all bidders. One policy implication is that it may pay the government to undertake a certain amount of exploratory activity on its own and make this information available to all bidders. In addition, it might be advisable to allow prospective bidders to undertake a certain amount of seismic and other exploratory activity before bidding on particular tracts, provided they turn the information over to the government and the government makes this information available to all bidders.

ROYALTY BIDDING. Although most oil-producing countries use royalties as one means of extracting rents for the government, experience with royalty bidding has been largely confined to that of the U.S. Department of the Interior in awarding OCS oil and gas leases.[14] (Royalty bidding also has been used in the Canadian provinces of Saskatchewan and Alberta.[15]) In contrast to bonus bidding, the petroleum company is exposed to less risk since it pays a royalty only if oil is found and produced. More firms are likely to bid since royalty bidding will attract smaller companies that are unable to pay large front-end bonuses. Less uncertainty will make for lower discount rates for calculating present values and a larger number of firms making bids is likely to result in higher bids.

The experience of the U.S. Department of the Interior in an OCS sale in October 1974, in which some tracts were selected for competitive bidding on the basis of royalty rates with a fixed bonus of $25 per acre while other tracts were offered on the basis of bonus bidding, showed that the number of bids per tract was substantially higher in the case of royalty bidding than in the case of bonus bidding.[16] However, these results do not prove that royalty bidding will maximize the government's economic rent as contrasted with bonus bidding, or that the tracts offered for bidding will be developed and explored more efficiently than under bonus bidding.

Unlike the case of bonus bidding where the tax is a sunk cost, royalties do affect the method of operation of the firm since they constitute a variable cost. Given a royalty, some discoveries will not be developed which might have been developed without a royalty, while those that are developed may be subject to less complete exhaustion of the resources and may be abandoned sooner than they would be in the absence of a royalty.[17] The reduced uncertainty which leads to more and higher bids on a royalty basis may be offset by a less efficient development of the discoveries.[18] In the 1974 Department of the Interior experiment, the winning OCS royalty bids ranged from a low of 51.8 percent to a high of 82.8 percent of the value of production.[19] Clearly, such a high tax on the value of production cannot help but affect adversely the volume of production compared with a low royalty or none. There can be a substantial differ-

[14]For a discussion of royalty bidding, see Stephen L. McDonald, *The Leasing of Federal Lands for Fossil Fuels Production* (Baltimore, Md., Johns Hopkins University Press for Resources for the Future, 1979), pp. 98–101; see also, U.S. Department of the Interior, *An Analysis of the Royalty Bidding Experiment in OCS Sale No. 36* (Washington, D.C., U.S. Department of the Interior, 1975).

[15]Dam, *Oil Resources*, p. 148.

[16]McDonald, *Leasing of Federal Lands*, pp. 99–100.
[17]Ibid. pp. 100–102.
[18]Ibid. p. 101.
[19]Interior, *Analysis of Royalty Bidding*, appendix D.

ence between the private value of marginal output and its social value. In addition, higher production raises the total amount of rent to be divided between the government and the producers.

Whether governments should choose royalty bidding versus bonus bidding may depend upon the amount of geological knowledge available on the oil lands in the areas put up for bid and on the degree of certainty with respect to the probability distribution of the outcome. In the case of risk contracts employed in bidding on offshore areas in Argentina and onshore areas in Guatemala where geological knowledge is minimal, there have been few bidders for contracts. In such cases royalty bidding would appear to be more effective than bonus bidding, at least until more discoveries are made. Once several discoveries are made and a substantial amount of geological knowledge is accumulated, many more companies will be attracted and bonus bidding may have the advantage of both increasing the share of the economic rent going to the government and of achieving a more efficient development of the oil resources discovered.[20]

PROFIT-SHARE BIDDING. So far as I am aware, bidding on the basis of the share of net profits going to the petroleum contractor has not been employed, but this approach has been examined in internal papers prepared in the U.S. Department of the Interior for application in leasing tracts on the outer continental shelf.[21] Under profit-share bidding, unlike the royalty system, the risks associated with cost uncertainties are shared between the government and the contractor and bids would be determined by the minimum share of the expected net returns that would make the investment attractive after allowance for commercial risk. Whether the lowest bid for a share of the profits going to the

contractor reflects the minimum share that would just make the investment attractive would depend on the bidder's assessment of the offers likely to be made by competing bidders.

There are several ways of defining profits, each of which may imply a different response on the part of producing companies. One concept of profits is "operating profits," which are equal to gross revenue minus operating costs. Operating costs usually exclude interest payments and capital charges.

The government's share of operating profits will have little effect on short-term efforts of petroleum companies to maximize net revenues unless the government's share is so large as to provide little incentive for efficiency. However, profit sharing without recovery of capital expenditures may result in a decision not to invest following exploration (which might not be the case if there were another fiscal arrangement) or discourage capital expansion for increasing output. Nevertheless, in contrast to a bonus bid, an equivalent amount of tax on profits paid to the government will result in a sharp difference in the internal rate of return on the investment and, hence, in the attractiveness of the investment. As is shown in table 4-2, the IRR on an investment with a million dollar cash bonus is only about one-third of that for the same investment in which a million dollars in profit sharing is paid to the government during the operating period. Clearly, a profit-sharing arrangement has substantial advantages over bonus bids in attracting risk capital, but profit-share bidding may not be the optimal choice for governments if there is no problem in attracting large numbers of bidders. In considering any tradeoff between the two systems, both the government's willingness to accept risk and its social rate of discount must be taken into account.

A second definition of profits that could be employed for profit-share bidding is revenue less operating costs and capital consumption allowances. This definition could be modified by allowing full recovery of all capital outlays before any profit sharing. An additional modification would be to allow full recovery of capital outlays plus a rate of interest on capital outlays equal to the cost of borrowing capital. This con-

[20]For a comparison of bonus and royalty bidding, see James D. Ramsey and John C. Sawhill, *Bidding on Oil Leases* (Greenwich, Conn., JAI Press, 1980) chapter 6.

[21]Marshall Rose and Donald Bieniewicz, "Policy Paper for Alternative OCS Leasing Arrangements," (Washington, D.C., U.S. Department of the Interior, January 1977) (unpublished).

Table 4-2. Examples of Effects of $1 Million Bonus and $1 Million in Taxes on Net Profits on Investor's Internal Rate of Return (IRR)
(millions of dollars)

Year	1	2	3	4	5	6	7	8	9	IRR
Net cash receipts with bonus in 1st year	-1.0	-0.2	-0.2	-0.2	0.6	0.6	0.6	0.6	0.6	12.9
Net cash receipts with 33% profits tax[a]	—	-0.2	-0.2	-0.2	0.4	0.4	0.4	0.4	0.4	37.3

[a]Tax is $0.2 million in each year of profitable operations, years 5–9.

cept of profits is closer to that of "pure economic rent," which allows for full economic costs.[22]

Profit-share bidding with profits defined to take into account the full economic costs of production, in contrast to a share of operating profits, further reduces the uncertainty of contractors, and thereby encourages entry and competition and lowers the rate of discount required by the bidders. However, it has the disadvantage of delaying payments to the host government since under full capital recovery the government would receive no payments until the investor recovered all of his capital plus interest.

A further argument advanced against profit-share bidding is that it is necessary for the lessor to monitor the financial records of the operating company. However, if the company is also subject to an income tax on its share of the profits, a profit-sharing arrangement would appear to require little or no additional financial monitoring on the part of the government.

PRICE BIDDING. Bidding on the basis of the price at which output is to be sold to the government or the price as a percentage of the world price has been employed by Argentina, Brazil, and Guatemala, among others. The economic effects of this type of bidding are similar to those of royalty bidding. They involve less risk to the contractor and are likely to attract more bidders than in the case of bonus bidding, but they may result in a decision not to develop marginal discoveries, or in an earlier termination of pro-

duction. They involve greater risk or uncertainty to the contractor than profitsharing since they are not affected by costs, and costs are one of the important variables subject to uncertainty. If costs prove to be higher than expected, there is no adjustment in the government's share of gross output. Finally, most companies prefer to receive a price for the oil they produce (or a share of the oil itself) based on the world market price. For example, petroleum company officials operating in Argentina and Colombia have expressed strong opposition to any form of government price fixing for petroleum.

EXPENDITURE BIDDING. Expenditure bidding, or bidding in terms of the number of wildcat wells the contractor commits himself to drill, has been used by Argentina and Guatemala, while in other cases there have been negotiations on "work programs" with the government requiring a minimum level of exploration activity.[23] Such negotiations have been carried on by the British government in granting concessions in the North Sea.[24] Contract terms in a number of countries, including Indonesia and the Philippines, require a certain amount of drilling and/or a certain level of exploration expenditures over a given period which are either stipulated in the model contract or negotiated.

Governments have sought to maximize drilling and other exploration activities by contractors in order to gain as much information as possible regarding the oil tracts, and to make sure the contractors engage in a minimum level of activity and do not simply hold the area as a

[22]The concept of profits could be made even closer to that of pure economic rent if economic rent were defined as any return in excess of that required to achieve an internal rate of return on invested capital that is equal to the rate of interest on borrowed capital.

[23]For a discussion of work-program bidding, see Jones et al. "Economic Issues," pp. 210–211.
[24]See Dam, *Oil Resources*, chapter 4.

reserve. Under competitive expenditure bidding, a company may commit itself to expenditures that may turn out to be wasteful. The company is usually required to pay the amount of any unrealized expenditures to the government (or forfeit a bond covering the commitment). The company will maximize its returns or minimize its losses if it adjusts its exploration expenditures on the basis of what it has learned as the exploration program continues. Although a government may want to negotiate a work program with a company in order to make sure that an appropriate amount of exploration is actually carried out over the period of the contract, there should be a provision that allows a company to be relieved of the work commitment on a particular tract after it has drilled one or two dry holes. Argentina's experience with expenditure bidding under the 1978 Risk Contract Law was not regarded as successful. Brazil also changed its model risk contract to permit companies to limit their expenditures to seismic surveys and to drilling one or two exploratory wells unless a discovery was made.

Fiscal Provisions in Nonbidding Arrangements for Contracts

Nonbidding arrangements are of two general types. First, provisions in standard contracts may be fixed by law or government decree, including the adoption of model contracts not subject to alteration by negotiation. Second, the provisions may be negotiated with the companies. As a rule, other conditions are fixed by law, even where requirements such as bonuses or gross revenue sharing are negotiable with individual companies or are subject to competitive bidding.

The principal difficulty with contract provisions fixed by law is that they cannot be altered with geological and other conditions, including the amount of information available to prospective contractors. This is obviously not the best way to maximize economic rents going to the government or to maximize the amount of exploration. In some cases governments have established specific contract provisions for certain areas, e.g., onshore versus offshore tracts, or special contract provisions based on the amount of exploration that has been done in an area, or proximity of the contract area to areas where oil has been found. This approach assigns to the government administrators the evaluation of alternative areas, but these evaluations may differ substantially from those of experienced petroleum companies.

There may be advantages in changing the provisions of future contracts on the basis of the experience of petroleum companies with older contracts. There have been a number of cases where governments have either tightened or liberalized their model contracts, depending upon how attractive they proved to be to prospective companies; or on the basis of the results of exploration activities of the initial contractors. Once important discoveries are made, companies are usually willing to sign contracts on terms more advantageous to the government. On the other hand, this strategy, which has been employed by several Latin American countries, may lead to an unwillingness of some companies to negotiate contracts until they see how contract terms evolve.

Individual negotiation of contract terms, while providing desired flexibility, may put the host government in a less favorable bargaining position than a system of competitive bidding. Also, prospective contractors may hold out for at least the most liberal terms that have been negotiated. Brazil has sought to bind companies that have negotiated contracts to secrecy regarding the financial provisions, and has refused to publish contracts. Nevertheless, at least one risk contract was (illegally) published in a Brazilian newspaper. In addition, it is almost impossible to keep anything secret in the community of petroleum companies. The author's experience in talking to officials of petroleum companies in Brazil suggests that they are well informed regarding all the contracts negotiated by their competitors.

Fixed and Sliding Scale Bonuses

A fixed bonus system has all of the disadvantages of a bonus bid, with the addition that the fixed bid may bear little relation to the potential economic rent of the petroleum field as perceived by the oil companies. If the fixed bid is too high there will be no bidders, while if it is too low, it may not capture the maximum share of the economic rent on the basis of the expected financial returns of the companies.

One form of bonus payment employed by Indonesia in a recent contract calls for the payment of $1 million at the time the contract is approved; $5 million on discovery of a commercially producible field; $7.6 million after recovery of the contractor's operating costs and investment credit; $17.1 million when daily production averages 50,000 barrels per day (bpd) for a period of 120 consecutive days; and $34.1 million after daily production averages 100,000 bpd for a period of 120 consecutive days. Although such an arrangement reduces uncertainty compared with, for example, a payment of a large bonus at the time of the negotiation of the contract, it is likely to prevent the development of a marginal field or the expansion of production beyond, say, 49,000 bpd. The rationale for a sliding scale bonus is that the larger the output the lower the cost per barrel, and, therefore, the larger the proportion of the total economic rent the host government can capture. However, a large progressive bonus obligation reduces the present value of marginal investments and may work against the most efficient development of a petroleum field.

Royalties

Nearly all petroleum-producing countries have established at least modest royalties. Royalties have the advantage of providing the host government with current income as soon as commercial production begins, in contrast to full reliance on profit-sharing arrangements. As discussed earlier, royalties have the disadvantage of preventing the most economical development of an oil field. One type that is frequently used is the "sliding-scale" royalty in which the royalty rate on marginal output rises as output from a field increases. The same result may be achieved by a provision which requires the company to pay an increasing proportion of the gross value

of output to the government as output expands. This device is found in contracts negotiated by Argentina, Brazil, and Guatemala, among others. The U.S. Department of the Interior experimented with a sliding-scale royalty in OCS Sale No. 43 on the Southeast Georgia Embayment in March 1978. In this sale the normal one-sixth royalty rose to a maximum of 50 percent as production from any tract rose above $1.5 million per quarter.[25] Of the 180 tracts offered that included this sliding-scale royalty, only 77 (or 43 percent) were leased, in contrast to a much higher proportion of leases sold for tracts in the same area without a sliding-scale royalty.

As in the case of the sliding-scale bonus, the rationale for the sliding-scale royalty or other tax based on gross output is that the larger the output, the larger the proportion of the total economic rent that the host government can capture. However, in terms of efficiency of development, the sliding scale royalty or a similar type of payment gets very low marks. To quote from a recent study, a sliding-scale royalty creates "contra-incentives to production, premature abandonment, retardation of intensive development investment, and loss of economic rent." Even in the case of tracts that are generally regarded to be highly productive and low risk, "it would be necessary to implement a sliding-royalty scale on each lease as it declined in productivity in order to insure that ultimate recovery would not be reduced. Otherwise the contra-incentive to intensive development investment would remain."[26] The sliding-scale royalty or a similar arrangement based on a proportion of gross output or gross revenue may not be a desirable method of maximizing economic rents for the host government.

Sale of Output to the Host Government at a Fixed Price

There are a few agreements, mainly those negotiated by Argentina and Colombia, that call for the delivery of oil produced by the contractor

[25]Jones et al., "Economic Issues," p. 208.
[26]Ibid. p. 208.

to the government at a negotiated price which is not related to the world price of petroleum. Such agreements were typical of the older style service contracts and have all the disadvantages of price fixing in a world of inflation. They tend to discourage production of marginal fields and provide no incentive for expanding output from existing wells through more costly recovery methods.

Indonesia and some other countries require that a certain proportion of the output of the contractors be delivered to the government at a low price for sale in the domestic market. This is a feature of virtually all Indonesian contracts that provide for payment of only 20 cents per barrel for oil after cost recovery. Where this arrangement represents a substantial proportion of total output—up to 25 percent of the contractor's share of the output in Indonesia—it reduces the present value of the contract and to a degree increases uncertainty for the contractor. Host governments may think they are subsidizing domestic consumption at the expense of the contractors, but this conclusion is clearly wrong. Inevitably they are reducing the government's economic rent, perhaps by even more than the subsidy to domestic consumers. In terms of fiscal policy, it often means that the government subsidizes relatively high-income nationals that use petroleum at the expense of the poorer citizens of the country.

Production-Sharing Contracts

Under production-sharing contracts (PSCs) in which the contractor makes all the outlays for exploration, development, and production while the output is split between the contractor and the state at the wellhead with no provision for recovery of any costs before the split, the share going to the state is similar to a royalty on gross output. This is true even where the state enterprise, as in the case of the Peruvian contracts prior to 1980, pays the income tax due the state out of the state agency's share of the output. Under this type of PSC, the share going to the government does not represent a tax on net profits after full cost recovery and is, therefore, not designed to tax pure economic rent. Not only

may the share taken by the government overtax or undertax economic rent, but the arrangement is not neutral with respect to investment decisions and may result in less than optimal development of the contract area.[27]

In some countries PSCs allow the contractor to deduct a certain percentage of the output, say, 40 percent, to cover his costs (cost oil) before sharing the remainder (profit oil) with the government. A cost oil allowance encourages the contractor to minimize his costs since his allowance does not decline with lower actual costs, thereby providing an incentive to increase efficiency. Also, the provision for cost oil in a PSC reduces the effective royalty level for any given profit oil split. Since operating costs as a percentage of sales revenue tend to decline as output rises, the expected NPV for a given percentage cost oil allowance is higher for large fields than for small ones. Therefore, for a given probability of an oil discovery, the contractor's expected NPV with a given cost oil allowance will be higher as the probability of finding a large field increases.

Charles Johnson has examined the relative advantages to the government of alternative arrangements for cost oil as a percentage of total revenue and for the profit oil split.[28] For example, is government revenue higher with a 40 percent cost oil allowance and an 80-20 profit split in favor of the government, or with a 30 percent cost oil allowance and a 70-30 profit split? He finds that no single PSC option will maximize the government's revenue from all exploration environments involving the size-frequency distribution of petroleum fields and the probability of discovering these fields. However, on the basis of an examination of several hypothetical cases, he finds that high-cost oil allowances combined with relatively high-profit oil shares tend to produce a greater expected NPV to the government than low-cost oil allowances combined with relatively low-profit oil splits.[29] Johnson found that in three of the four

[27]Craig Emerson, "Taxing Natural Resource Projects," *Natural Resource Forum*, April 1980, pp. 123–145.
[28]Johnson, "Establishing Effective Production-Sharing," pp. 129–141.
[29]Ibid. p. 136.

hypothetical exploration environments examined, a 40 percent cost oil allowance and an 80-20 split in favor of the government would be more advantageous to the government than a lower cost oil allowance and a profit oil split less favorable to the government, say, 70-30. However, the high-cost allowance, lower profit split arrangement places a relatively high fiscal burden on small fields so that they might not be developed by the contractor. This could lower the risk-corrected NPV of the entire investment to the contractor and, therefore, reduce its attractiveness.

Windfall Profits Taxes

In addition to imposing normal income taxes on net profits, some governments have imposed so-called "windfall profits" taxes on petroleum producers in order to capture a larger share of the economic rents. Windfall profits taxes take many forms, but most involve higher rates of taxation on accounting rates of return above a certain percentage. Since a contractor's expected NPV depends heavily on the possibility of discovering a large field that will yield high returns, the effect of a windfall profits tax is to reduce substantially the expected NPV from a potential investment. If a government were to attempt to tax windfall profits at 90 percent, it is unlikely that any high-risk investments would be made.

Accounting profits vary widely over time and do not constitute a proper measure of economic rent since they do not reflect full economic cost. The only proper measure of pure economic rent is the surplus over the net return to the investor that yields an IRR equal to the opportunity cost of capital. Moreover, if an investment is to be made, this IRR must be adjusted for risk.

The "Resource Rent Tax" or DCF "Trigger Tax"

One means of avoiding some of the difficulties with an excess profits tax on the accounting rate of return is provided by the so-called "resource rent tax" or the DCF "trigger tax."

This tax system, which was introduced by Ross Garnaut and Anthony Clunies-Ross,[30] is illustrated in table 4-3. The resource rent tax is levied only on profits in any year that are in excess of the amount necessary to yield a specified IRR on all capital expenditures. In the example given in table 4-3, the IRR is 18 percent and this rate is not realized until the eighth year of operation, at which point a resource rent tax of 60 percent applies to net cash receipts in that year that are in excess of the amount necessary to achieve the 18 percent IRR on total capital invested since the initial investment was made.

A major difficulty with the resource rent tax when applied to a prospective investment in an unexplored petroleum field is that it would reduce the expected NPV on a discovery of a large field and, hence, the combined NPV for the investment. Moreover, it is the possibility of discovering a large petroleum field, even though the probability may be very low, that is often most attractive to petroleum firms. Only if the government knew the minimum acceptable (risk-corrected) NPV of a petroleum firm with respect to a prospective investment and established a resource rent tax on net profits in excess of the corresponding IRR, could it employ a resource rent tax without risking the possibility that the petroleum investment would be rejected.

As the sole source of revenue to the government, the resource rent tax has certain drawbacks which are recognized by its authors. First, if this system is applied to foreign investors whose home governments provide foreign tax credits against taxable dividends transferred to the home country, such tax credits would not be available until a resource rent tax was actually paid. (This would be true for the U.S. investors, but not for the investors of certain other countries.) Second, the host government would receive no profits tax rev-

[30]Ross Garnaut and Anthony Clunies-Ross, "Uncertainty, Risk and Taxing of Natural Resource Projects," *Economic Journal* (June 1975) p. 287; see also, Raymond F. Mikesell, *The World Copper Industry* (Baltimore, Md., Johns Hopkins University Press for Resources for the Future, 1979) pp. 295–297.

Table 4-3. Hypothetical Example of Tax on Accumulated Present Value
(DCF trigger tax)

	Net cash receipts	A_t Accumulated present value at 18%	$T(A_t)$ Tax on returns over 18% threshold at tax rate (T) of 60%
Expenditures in the first year of capital outlays	−130	−130	
Expenditures in second year of capital outlays	−100	−253	
Expenditures in third year of capital outlays	−100	−399	
Year of operation			
1st	100	−371	
2nd	100	−338	
3rd	100	−299	
4th	100	−253	
5th	100	−199	
6th	100	−135	
7th	100	− 59	
8th	100	30	18
9th	100	—	60
10th	− 50	− 50	0
11th	100	41	25
12th	100	—	60

Source: Based on an example by Ross Garnaut and Anthony Clunies-Ross in "Uncertainty, Risk Aversion and the Taxing of Natural Resource Projects," *Economic Journal* (June 1975) p. 287.

Note: In the second year of capital outlays, $A_t = -130(1.18) - 100 = -253$. In the third year of capital outlays, $A_t = -253(1.18) - 100 = -399$. In the first year of operations, $A_t = -399(1.18) + 100 = -371$. And so on. In the eleventh year of operations, $A_t = -50(1.18) + 100 = 41$.

enue until the resource tax was applied. Since an investor might be earning fairly high accounting rates of return for a number of years before the resource tax cuts in, there might be political opposition to the arrangement within the host country. This problem might be dealt with by imposing a lower tax on accounting profits in addition to the resource rent tax.[31]

The first country to employ the resource rent tax in petroleum was Papua New Guinea. The basic PNG corporate tax rate is 50 percent on net profits and the rate of accumulation, or IRR, is 27 percent; any net cash flow in excess of the accumulated present value at 27 percent is taxed at a rate of 50 percent. This means that the marginal tax rate is about 75 percent. Recently the resource rent tax concept has been employed in petroleum agreements negotiated by Liberia, Equatorial Guinea, and Tanzania.

[31]For a discussion of the drawbacks of the resource rent tax, see Keith F. Palmer, "Mineral Taxation Policies in Developing Countries: An Application of the Resource Rent Tax," *Staff Papers* (Washington, D.C., International Monetary Fund, September 30, 1980) pp. 517–542.

Capital Recovery Provisions

The rate of capital recovery permitted in the calculation of taxable income is exceedingly important to potential private investors. The higher the rate of capital recovery, the higher will be the estimated IRR on an investment and the lower will be the perception of risk. A higher rate of capital recovery may also encourage additional investment by a petroleum firm since firms are generally more willing to reinvest cash flow from a foreign project than to finance additional investment from abroad. The resource rent tax discussed above provides for full capital recovery before any tax is levied, but full capital recovery can be achieved by 100 percent depreciation and amortization of nondepreciable assets before the application of any tax on revenue. This is the case with the U.K. North Sea contracts.

From the standpoint of the host government, one disadvantage of a high rate of capital recovery is that it delays government income from the project. This is true regardless of how the contract is otherwise structured. Governments must balance the higher present value of early returns against the potentially higher rents that might be obtained after rapid capital recovery. The potentially higher rents would be derived from the lower discount rates investors would use in calculating the present value of investments, the larger number of firms that would be attracted to bidding on or negotiating contracts, and the encouragement to reinvest the higher cash flow from capital recovery.

The Effects of Alternative Fiscal Systems on the Development of New Petroleum Fields of Differing Quality

From the standpoint of the volume of reserves (and, hence, production potential in terms of barrels per day) and capital costs in terms of costs per barrel recovered, the same fiscal system or combination of fiscal arrangements, such as bonus payments, income taxes, and production sharing, may have substantially different impacts on the after-tax NPV for petroleum fields of varying quality. Petroleum regions of the world may be classified on the basis of exploration and production experience, in terms of average reserves, volume, and capital costs per barrel. For example, Argentina's onshore fields tend to be low volume and high cost, while the Middle Eastern fields tend to be high volume and low cost. Capital costs vary with the average depth of the wells, the physical terrain (e.g., mountainous, jungle, or plain), and distance from ports, as well as with the output of the wells. Offshore fields in deep water have a higher capital cost than onshore fields. Philippine offshore fields thus far have been shown to have a low volume and a high capital cost.

The petroleum fields of most OIDCs tend to be characterized by relatively low volume reserves and high capital costs per barrel compared with Middle Eastern fields.[32] Therefore, if the fiscal systems tend to discriminate against

low volume/high cost fields in their effects on expected NPV or IRR, they will reduce not only the attractiveness of exploration in countries where the vast majority of the fields are characterized by low volume and/or high cost, but also the attractiveness of field development by contractors after the area has been explored. An examination of the effects of different types of fiscal systems on fields of differing quality is complicated by the fact that most petroleum-producing countries have a mixture of fiscal arrangements. However, certain types of fiscal systems, such as production-sharing, royalties, and bonus payments, tend to be regressive in the sense that the government's share of net profits is significantly larger on higher quality fields (e.g., low cost/high volume) than on lower quality fields (high cost/low volume). On the other hand, progressive income taxes, the resource rent tax, and rapid cost recovery tend to be more progressive in their effects on the government's share of net profits, and therefore provide greater encouragement for the development of high-cost/low-volume fields. Since the before-tax expected IRR is likely to be substantially lower for low quality than for high quality fields, a regressive or even a proportional tax system may result in an expected IRR for a high-cost/low-volume field that is unacceptable to investors. At the same time it may provide a relatively high IRR for low-cost/high-volume fields which may have a discovery probability of close to zero in a particular region. Therefore, a regressive fiscal system will tend to reduce both the expected IRR of the investment and the risk-corrected IRR.

Two petroleum economists have simulated the effects of fiscal systems in recent contracts negotiated by five countries on the government's share of resources generated and on real internal rates of return from four hypothetical and identical oil fields.[33]

[32]According to Richard Nehring, only two OIDCs, India and Colombia, have any known giant fields (over 500 billion barrels). Richard Nehring, *Giant Oil Fields and World Oil Resources* (Santa Monica, Calif., Rand, June 1978) pp. 32–33.

[33]The results of the study are summarized by Alexander G. Kemp and David Rose in "Four Oil Fiscal Systems in Need of an Overhaul," *Financial Times Energy Economist*, April 1982; see also Kemp and Rose, *Investment in Oil Exploration and Development: A Comparative Study of the Effects of Taxation* (Aberdeen, Scotland, University of Aberdeen, North Sea Study Occasional Papers, November 1982)

The four hypothetical (new fields) are characterized as low cost/high volume, medium cost/medium volume, high cost/medium volume, and high cost/low volume.[34] The fiscal systems examined were those of Egypt, Indonesia, Malaysia, Nigeria, and Papua New Guinea. In all cases the simulations were on a project lifetime basis and were conducted on both a nominal and a real return basis, and assumed constant real oil prices. The price of oil in the base case was $33.80 per barrel, and in the calculation of real IRR both the price of oil and capital and operating costs were escalated at 9 percent per year. In the absence of government taxes, all four hypothetical fields yield a positive NPV at a 15 percent (real) rate of discount.

From table 4-4 it may be seen that in PNG the government's take (nominal basis) as a percent of resources generated (total revenues minus all capital and revenue costs) declines progressively from the low-cost/high-volume field to the high-cost/low-volume field; in all four hypothetical fields the (real) IRR is significantly above the minimum acceptable rate of 15 percent for most petroleum investments. In Egypt and Indonesia, the IRRs for both the low cost/high volume and the medium cost/medium volume are significantly above the minimum acceptable rate, but the IRRs for the high-cost fields are not. In Malaysia and Nigeria, IRRs on the medium-cost/medium-volume fields are only marginally above the minimum acceptable rate of return, but in both cases the low-cost high-volume fields are significantly above the minimum rate of return.

The PNG fiscal system for petroleum is characterized by a very low royalty rate of 1.25 percent and a net profits tax plus a resource rent tax. The Egyptian, Indonesian, and Malaysian systems employ production-sharing arrangements that provide that a certain quantity of oil is first allowed for the recovery of exploration

and production costs while the remainder is split between the private and government oil companies.[35] The Egyptian and Indonesian systems are characterized by more generous cost recovery arrangements than in the Malaysian system, and in the latter system the private oil company has to pay an income tax of 45 percent on its share of crude production plus a further export duty of 25 percent. There is also a royalty of 10 percent based on an official price somewhat below world market levels. The Indonesian system does not provide for a royalty, but does provide for a net corporate profits tax and there are also signature and production bonuses. The Nigerian system provides for a posted price set above the market price, and the royalty, varying from 16⅔ to 20 percent, is based on this price. Participation by the Nigerian government in foreign petroleum companies is 60 percent, with the government paying its share of development costs, but exploration costs are carried by the company and repaid over three years once the field is declared commercial.

An income tax tends to be more progressive (or less regressive) in its impact on the proportion of government take of resources generated and on the IRR than most other fiscal instruments such as royalties and production sharing. The degree of progressivity of an income tax depends upon the provisions for capital recovery and whether there are progressive tax rates. Indonesia and Malaysia both employ a combination of production sharing and a corporate income tax. In Indonesia, the share of the corporate income tax in total government take is 18.9 percent for the low-cost/high-volume hypothetical oil field, rising to 21.8 percent for the high-cost/low-volume field; while in Malaysia the share of the income tax in the total government take is 9.3 percent for a low-cost/high-volume field, declining to 1.9 percent for the high-cost/low-volume field. The higher share of the income tax in total government take helps to explain the lower degree of regressiveness of the Indonesian tax system compared with the Malaysian tax system, although both are actually regressive.

---

for assumptions and a description of the model employed in the simulations.

[34] A high-volume field is defined as one with a peak production of 200,000 bpd; a medium-volume field, 70,000 bpd; and a low-volume field, 30,000 bpd. Low-cost = $3,000 per peak daily barrel recovered; medium-cost = $12,000; and high-cost = $25,000.

[35] Details of the Indonesia, Malaysia, and PNG systems are given in subsequent chapters.

Table 4-4. Results of Simulations of Operation of Fiscal Systems of Five Countries for Four Hypothetical Oil Fields

	Low cost/ high vol.	Med. cost/ med. vol.	High cost/ med. vol.	High cost/ low vol.
Egypt				
Total govt. take as percent of resources generated[a]	87.5	83.4	82.2	81.2
IRR (real values)	55.8	24.3	8.2	5.4
Indonesia				
Total govt. take as percent of resources generated[a]	81.2	81.3	82.7	83.4
IRR (real values)	109.9	35.4	9.3	6.5
Malaysia				
Total govt. take as percent of resources generated[a]	88.2	88.3	93.3	93.4
IRR (real values)	70.2	15.8	neg.	neg.
Nigeria				
Total govt. take as percent of resources generated[a]	95.9	95.7	94.6	93.4
IRR (real values)	66.2	18.8	3.4	0.9
PNG				
Total govt. take as percent of resources generated[a]	80.7	79.4	74.1	70.7
IRR (real values)	122.1	43.6	22.2	21.0

Source: Kemp and Rose. "Four Oil Fiscal Systems in Need of an Overhaul." *Financial Times Energy Economist.* April 1982.
[a]Nominal basis. Resources generated are total revenues minus all capital and operating costs.

In the PNG tax system, nearly all the government's income is derived from the income tax, including the progressive resource rent tax, so that the government's take as a percentage of resources generated is significantly lower for the high-cost/low-volume field than for the low-cost/high-volume field. The relatively high proportion of the income tax in the government's take in the Nigerian system also explains the slightly progressive impact of the Nigerian system.

The relative progressivity of the nominal government take in the Egyptian fiscal system arises from the provision requiring companies to return to the government all cost recovery oil not actually needed to recover costs. In the more profitable fields, e.g., low cost/high volume, proportionately more cost recovery oil must be returned to the government over the life of the project than in the less profitable fields, e.g., high cost/low volume. In general, however, it is clear that, with the exception of PNG, all of the systems discussed have a severe impact on the high-cost/low-volume fields, so that their development tends to be relatively unattractive to petroleum companies.

The obvious conclusion from the simulations undertaken by Kemp and Rose is that tax progressivity is likely to increase government revenues by raising the investor's NPV of the lower quality fields to acceptable levels and therefore raising the combined expected NPV for the exploration area, especially where the probability of discovering low cost/high volume fields is very low. By rendering low quality fields profitable, a progressive tax system will also reduce the risk perception of the investors (as measured by the coefficient of variation of the outcomes) and, therefore, raise the risk-corrected IRR for the investment.[36]

Government Equity Participation

There is wide variety in arrangements for government equity participation, each of which affects the division of economic rents, the attractiveness of exploration and development, and the efficiency of development.

[36]Kemp and Rose have also undertaken simulations for four hypothetical fields employing the fiscal systems of several developed countries, including the U.K., Norway, Australia, and several systems employed for the U.S. outer continental shelf and by the state of Alaska. The results in terms of the progressivity of the tax systems conform with the findings with respect to the fiscal systems of the five developing countries described above. See Kemp and Rose, *Investment in Oil Exploration.*

The Pure Joint Venture

A pure joint venture exists when both the host government (or a government agency) and the private petroleum company share fully in the risks of exploration and development and in the returns from the investment in proportion to their equity participation. The joint venture may be subject to payment of royalties and to corporate income taxes, but such taxes are imposed on the joint venture rather than on the private company alone. Pure joint ventures are virtually unknown in petroleum, although Indonesia employs a combination of a joint venture and a production-sharing agreement between Pertamina and private petroleum companies.

The government usually is better off being in partnership with an experienced international petroleum company rather than undertaking a project on its own. However, the government may be less well off than it would be if it permitted a large company to take all of the risks, since a large petroleum company is able to reduce risks by making a number of investments in different countries and regions throughout the world. An international petroleum company is a professional risk taker and portfolio manager. Its cost of risk bearing will be considerably less than that of the government of the average developing country.

Joint Venture with Unequal Risk Sharing

In the typical joint-venture arrangement, the private company bears the cost (and risk) of exploration until commercially producible oil is discovered. If both partners agree to carry on development of the field, the private company is compensated for the exploration outlays, either by direct reimbursement or by recovery from future output. This arrangement is called a "carry." In some joint ventures the government agrees to share the cost of development in proportion to its equity interest, while in other cases the state is "carried through development" and does not share in costs until commercial production begins. Although development is less risky than exploration, if the private partner finances development, he continues to bear both cost and

risk for the entire venture, but receives profits only in proportion to his equity interest. In most cases reimbursement for exploration and development costs does not include interest on capital investment, so that such costs must be entered into the calculation of expected NPV by the petroleum company negotiating the joint venture.

Regardless of how the joint venture is structured from the standpoint of risk sharing and the method of recovering capital outlays, the private contractor will calculate the NPV of the contract on the basis of the probability distribution of possible returns to his investment and employ a discount rate appropriate to the degree of uncertainty. However, our concern is whether the joint venture is more or less advantageous for the state compared with 100 percent private ownership from the standpoint of maximizing government revenue. We must assume, of course, that in the absence of a joint-venture arrangement the state will seek to negotiate higher royalties or income taxes (or both) to the degree that it is able to maximize its share of the economic rent and still attract the investor. Hence, we are concerned with alternative methods of extracting rent.

Let us first consider the case in which the private partner assumes all the risk and capital outlays in both the exploration and development stages in a 50-50 joint venture with a state agency and the private partner is fully compensated for the capital outlays before there are any profits available for distribution. We also assume that no taxes are levied on the joint venture. We may compare this arrangement with a 100 percent privately owned company undertaking the same project, but subject to a 50 percent tax on net profits after full recovery of capital. Assuming no other taxes under either scenario, will the economic rent going to the host government be the same in both cases? Not necessarily. For one thing, a portion of the net profits will be required to make additional investments for expanding production. Taxes on net profits of the 100 percent privately owned company will be higher than the amount available for distribution to the state agency in the 50-50 joint venture. Perhaps even more important, if the private petroleum company is able to offset the 50 percent corporate income tax against its tax obligations to its home country, but not the 50 percent share accruing to the state in the joint venture (as is the case with a U.S. firm operating abroad), the present value of a 100 percent privately owned operation will be substantially greater to the petroleum company than the present value of its share of a joint venture, while the risks in both cases will be the same.

If in the joint venture the state accepts more of the risk by contributing its share of the cost of development, the present value of the project to the company will be higher and there will be a possibility for the state to extract more rent, say, by demanding a larger share of the equity. However, risk is not costless and the social cost of risk to the state could well be higher than the cost to a private company with well-diversified investments throughout the world.

There are a number of other scenarios that we could explore, but there does not appear to be a strong case for any financial advantage to the state from a joint venture compared with 100 percent foreign-private ownership. From the standpoint of the attractiveness of the investment, some companies may reject them outright while others may regard certain forms of joint ventures as advantageous because they reduce risks. There may also be cases where the private contractor believes that joint ventures provide some protection against expropriation or contract violations.

Conclusion

This chapter has analyzed the revenue implications for the government and the effects on private company investment decisions for a variety of methods of sharing revenue derived from petroleum operations. Nearly all of these methods are represented in one or more of the case studies in part II. None may be regarded as absolutely undesirable from the standpoint of the host country, although some are clearly superior to others, depending upon such conditions as knowledge of the geology of the exploration area or the nature of the competition for contracts, and upon the objectives of the host country. What I have sought to show is that there are tradeoffs in the evaluation of any of these arrangements in comparison with alternatives.

APPENDIX 4-A
CALCULATION OF RISK-CORRECTED INTERNAL RATE OF RETURN*

The formula for calculating the risk-corrected internal rate of return discussed in the text can be written as follows:

$$E = P_s (F_s - T_s) + P_L (F_L - T_L)$$
$$V = P_s (F_s - T_s)^2 + P_L (F_L - T_L)^2 - E^2$$
$$S = \sqrt{V}$$

where:

E = expected present value from the two ventures net of taxes

F_s = pretax present value from small field

F_L = pretax present value from large field

T_s = present value of tax burden on small field

T_L = present value of tax burden on large field

V = variance of the portfolio of the two ventures

S = standard deviation of portfolio of two ventures

P_s = probability of success on small field

P_L = probability of success on large field

The expression for E can be regarded as the "long-run average" present value of the portfolio of two ventures. It is the present value the decision maker could "expect" on average if the investment in two ventures were repeated a large number of times. The expression for V can be regarded as a measure of how much the present value of any *particular* portfolio of two ventures may depart from the long-run average—it measures "variability." The square root of V, i.e., S, is the standard deviation. It can be taken as measuring the degree to which the present value of the two ventures is "spread" about the long-run average. V (or S) can, thus, be regarded as measures of risk or, more properly, as the decision maker's risk perception of any portfolio of two ventures.

————
*The author is indebted to Dr. M. A. Grove of the Department of Economics, University of Oregon, for the preparation of this appendix.

Using the numerical example in the text, it is easy to see that the decision maker's risk perception, V, depends critically on the tax system assumed to be in effect. We can calculate the following:

Tax system A	Tax system B
$E = 100$	$E = 100$
$V = 100,000$	$V = 34,000$
$S = 316$ (approximate)	$S = 184$ (approximate)

Note that the decision maker's risk perception (as measured by V) under tax system B is 34 percent of his risk perception under tax system A.

Since E measures *long-run* success, and V measures the variability in any particular portfolio of two ventures, a criterion for choosing ventures in any *short-run* setting would be one where ventures are chosen to make E as large as possible and V as small as possible. A criterion that includes these two goals of performance in a single statement is given by the formula

$$u = E - aV$$

where a is a prespecified (positive) constant. Values of a larger than one mean that risk (V) has a great deal of importance to the decision maker: he "discounts" long-run success heavily for risk. Values of less than one mean that risk has less importance to the decision maker. Thus, a measures the decision maker's "risk aversion."

The decision maker's risk-corrected IRR can be measured along the same lines as those discussed above, i.e., the risk-corrected IRR, e^*, can be written as

$$e^* = e - aV$$

where

e = the expected IRR
V = the variance of IRR
a = risk aversion (discount)

To see how this formula works, suppose that e

is 40 percent, and that V under tax systems A and B is 0.66 and 0.34 respectively. Then if a = 0.35, the risk corrected IRR under tax systems A and B is 16.9 and 28.1 percent respectively.

APPENDIX 4-B
A HYPOTHETICAL EXAMPLE OF BONUS BIDDING AND THE EFFECTS ON GOVERNMENT REVENUE

The effects of a bonus bid system on government revenue can be illustrated by a hypothetical example of a risk-neutral petroleum firm.[37] The following assumptions are used in this example:

a. First year—high-risk exploration outlay of $20 million with a 0.25 success probability

b. Second year—low-risk exploration and development expenditures of $100 million

c. Third year—low-risk development expenditures of $70 million

d. Annual sales revenue over a ten-year period (assumed to be the life of the field) of $180 million

e. Annual operating cost of $27 million during each of the ten years of operation, and

f. The company's IRR target is 20 percent over a ten-year operating period.

Appendix table 4-B-1 shows the annual cash outflow and inflow to the petroleum company over the thirteen-year period. For the first year the $20 million high-risk exploration costs trans-

Table 4-B-1. Net Present Value of Petroleum Project at 20 Percent Discount
(millions of dollars)

Year	Cash flow (− = outflow)	Cash flow (− = outflow)
1	$− 80	$− 80
2	− 100	− 100
3	− 70	− 70
4	153	73.3
5	153	73.3
6	153	73.3
7	153	73.3
8	153	73.3
9	153	73.3
10	153	73.3
11	153	73.3
12	153	73.3
13	153	73.3
Net present value	190 (before tax)	0 (with 52 percent net operating profits tax)

late into an $80 million outlay when adjusted for the 0.25 probability coefficient. Annual cash inflow equals $180 million in sales revenue less $27 million operating expenditures, or $153 million.

In the absence of any bonus, royalty, or government taxes, the net present value of the project to the company at a 20 percent rate of discount is $190 million. Under conditions of competitive bidding, the highest bid that can be expected (¼ × $190 million) is $48 million,

[37]See Hayne E. Leland, "Technical Appendix" to "An Economic Evaluation of Alternative Leasing Systems: A Report to the State of Alaska and to the Department of Natural Resources," (mimeo), Department of Natural Resources, State of Alaska, October 9, 1979. In this technical appendix Leland presents a model for comparing bonus bids for petroleum leases with profit-sharing arrangements. The above analysis draws on this model.

assuming all the bidders employ the same rate of discount and economic evaluation of the project. [The same probability coefficient (0.25) must be assigned to the bonus bid as that assigned to the high-risk exploration outlay.]

Given the same capital outlays and without a bonus bid, the company could realize a 20 percent IRR if the annual net operating profit were only $73.3 million instead of $153 million (table 4-B-1). Therefore the government could tax operating profits at 52 percent, yielding $79.6 million per year, or $796 million for the entire ten-year period. Hence, if the government's social discount rate were zero, it could receive *every year* an amount of revenue which is higher than the maximum bonus it could expect to receive under competitive bidding. If we assume that the government has the same discount rate as the company, namely, 20 percent, the present value to the government of a series of payments equal to $79.6 million beginning with the first year of operations (the fourth year of the project) would be $190 million compared with a maximum bonus receipt of $48 million. If the government's discount rate were 15 percent, the present value of tax revenues from a 52 percent tax on net operating profits would be $263 million; if the government's discount rate is 12 percent, the present value of tax receipts would rise to $321 million.

Instead of a tax on net operating profits of 52 percent, the government might levy a royalty of 44 percent on sales revenue ($180 million per year) and realize the same return. However, a royalty of 44 percent would constitute a substantial increase in marginal cost and is likely to reduce the output of the field. (A 44-56 production-sharing arrangement without provision for cost recovery would give identical results.)

The government might also employ a net profits tax instead of a bonus. Assuming a ten-year depreciation schedule of 10 percent of the *actual* exploration and development outlays of $190 million, taxable income each year would be $134 million ($153 million − $19 million). The maximum net profits tax that the government could impose on taxable income would be 59.4 percent. The government's revenue under the assumptions given above would be the same whether it imposed a 52 percent tax on net operating profits, a 44 percent royalty, or a 59 percent tax on net profits.

Taxing either net operating profits or net profits (assuming no additional capital outlays) would have an advantage over royalties, but if additional capital outlays were required, the tax on net profits would have an advantage over a tax on net operating profits. The reason is that, in the absence of depreciation, development expenditures on marginal projects might not be made.

It can be shown that if the government's rate of discount is lower than that of the company, the present value of the government's revenue would be maximized by accelerated depreciation and a net profits tax. With accelerated depreciation, the company would pay no tax during the first year of operations and a slightly reduced tax during the second year. This would increase the company's net present value substantially, which in turn would permit a substantial increase in the profits tax rate and, hence, in the return to the government over the ten-year period. The net profits tax could be raised from 59 percent to over 70 percent with accelerated depreciation, and the present value of government revenues at a 12 percent rate of discount would be increased substantially over the case with depreciation at 10 percent annually. However, accelerated depreciation would not benefit the government if its own rate of discount were equal to or higher than that of the company.

II

CASE STUDIES OF MODERN CONTRACTUAL ARRANGEMENTS

to Indonesia but their activities were hampered by Indonesia's struggle for independence, which ended in the transfer of sovereignty to the Indonesian government in December 1949. The Indonesian Parliament passed a motion in August 1951 which prohibited the granting of new concessions while the government was developing its petroleum policy. The new policy was finally embodied in Oil and Mining Law No. 44 which was signed by President Sukarno in 1960. Law No. 44 established the principle that "Oil and natural gas mining is only conducted by the State and the State company is authorized to engage in oil mining on behalf of the State."[2] Under this law, the existing concession agreements were, after considerable negotiation, transformed into "contracts of work" under which the foreign operators became contractors to state enterprises. However, for a time the new contracts did not change the conditions under which the former concession holders operated, and political conditions in Indonesia discouraged new investment. Shell sold its Indonesian operations to Permina and withdrew from the country in 1965,[3] but the two American companies have continued to operate under their contracts of work negotiated in 1963. Caltex's contract expires in 1983 as does one of Stanvac's contracts, with the other expiring in 1993. In 1980 over 45 percent of Indonesia's crude output was produced by Caltex.

The basic petroleum policies of Indonesia as they affected operations of foreign petroleum producers were not established until the overthrow of President Sukarno and his replacement in 1966 by General Suharto, who served as acting president until he was elected president in March 1968. Meanwhile, three state oil companies—Permina, Pertamine, and Permigan—were created, and in 1968 the first two enterprises were consolidated into a single entity, PN Pertamina.

The instrument adopted for carrying out Indonesia's petroleum policy set forth in Law No. 44 of 1960 is the production-sharing contract,

[2]Ibid., p. 8
[3]Shell signed a production-sharing contract for a tract in Irian Jaya (Indonesia) December 12, 1979.

which was devised by Dr. Ibnu Sutowo, later president of Pertamina. The first such contract was signed between Permina (which later became Pertamina) and the Independent Indonesian American Petroleum Company (Iiapco) in 1966. This contract embodied the basic principles of the production-sharing contracts that have been negotiated with foreign companies to the present. They are as follows:

1. Pertamina has responsibility for the management of petroleum operations and the contractor is responsible to Pertamina for the execution of such operations in accordance with provisions of the contract.
2. The contractor provides all financing and technology required for the operations and bears the risk of production costs.
3. During the term of the contract, total production after allowance for operating costs is divided between Pertamina and the contractor in accordance with provisions of the contract.
4. The contractor must submit annual work programs and budgets for scrutiny and approval by Pertamina.

Aside from the basic principles, the provisions of the contracts have changed over time, particularly with respect to the production-sharing ratio, the method of calculating production costs, and the arrangements for paying Indonesian taxes. Under the first production-sharing contracts with Iiapco, a maximum of 40 percent of the annual output of petroleum could be recovered by the foreign company to meet its preproduction and production costs. After deducting production costs, the balance of the petroleum produced was split 65 percent to the state company and 35 percent to the foreign company. The Indonesian income tax for which the foreign company was liable was paid by the state company. The contractor was free to market its share of the output without repatriating the foreign exchange from the sale. The contractor also normally marketed the oil representing the operating costs and Pertamina's share of the oil, and the price was established as the net realized price f.o.b. Indonesia for crude sold to third parties.

As new production-sharing contracts were signed, additional features were added to the contracts, such as payment of signature and production bonuses, differential splits in output based on the number of bpd produced, and the requirement by the contractor to supply a portion of its share to the domestic market. The 1968 contract between Pertamina and Phillips/Superior required the contractor to refine in Indonesia up to 10 percent of its share of the crude by setting up a refinery, or refineries, in Indonesia, provided the contractor's share of the crude exceeded 200,000 bpd and that such refineries would be economically feasible. Under the same contract, if production reached an average level of 100,000 bpd for 180 consecutive days, the split was changed to 67.5 percent for Pertamina and 32.5 percent for the contractor. Also, the contractor agreed to pay Pertamina $500,000 within 15 days of the effective date of the contract and $2 million after production exceeded 100,000 bpd for 180 consecutive days. The contractor also agreed to deliver a portion of its share of the output to the domestic market based on the ratio of total amount of crude produced from the contract area to the total amount of crude produced in Indonesia, times the amount of crude consumed in Indonesia, with a limit of 25 percent of the contractor's percentage share of the crude production from the contract area. The price of oil sold to the domestic market must be under 20 cents per barrel.[4]

Changes in Contract Terms Following the 1973–74 Rise in Petroleum Prices

Following the sharp rise in world petroleum prices in the winter of 1973–74, the Indonesian government demanded that the contracts be renegotiated to increase Pertamina's share of the output. Under agreements negotiated by Pertamina with Caltex and with Atlantic Richfield in March 1974, and later applied to other contracts, the split provided in the original con-

tracts, usually 65 to 35, applied only to the first $5 per barrel received by the contractors. The $5 per barrel price was adjusted with the rise in the UN Commodity Index since 1973. On revenue in excess of the base price, Pertamina received an 85 percent share on the first 150,000 bpd of production; 90 percent on production from 150,000 to 250,000 bpd; and 95 percent above 250,000 bpd.[5] Depending upon the price, the new split worked out to about 85–15 in favor of Pertamina. This unilateral action was, of course, resented by the companies, but it apparently did not reduce their profits to the point at which they were unwilling to negotiate additional contracts.

Following the dismissal of Dr. Ibnu Sutowo as president of Pertamina in March 1976, as a consequence of the state company's chaotic financial affairs,[6] the Indonesian government again violated its contracts with the petroleum companies. In April 1976 the government reduced the estimated profit margin of the largest producer, Caltex, by $1 per barrel (or about $300 million per year) by renegotiating its work contract. In the case of companies operating under production-sharing contracts, the government raised its 65 percent share of oil left after cost recovery to 85 percent, and increased the period over which companies could recover their costs from seven to fourteen years. Since the producing life of many of the fields was only five to seven years, this action made it impossible for some companies to recover costs and finance further exploratory drilling.

The profit margin of the production-sharing contractors was reduced by $1.25 to $1.00 per barrel. On top of these developments, in April 1976 the U.S. Internal Revenue Service ruled that the output shares of foreign governments under production-sharing contracts were to be treated as royalty payments and not allowed (as had been the case) as credits against U.S. tax obligations. As a consequence, petroleum

---

[4]"Summary of Indonesian Oil Contracts," Pertamina, Jakarta, Indonesia, 1972, pp. 29–32.

[5]The formula differed depending upon whether the contractor was operating under a contract of work, as was the case for Caltex and Stanvac, or under a production-sharing contract.

[6]Sutowo was replaced as president of Pertamina by General Piet Haryono.

companies sharply reduced exploration commitments in Indonesia in 1976 and 1977.[7]

In response to the companies' reaction to the unilateral changes in their contracts, the government offered several financial incentives to explore for new oil and to undertake enhanced recovery operations. The government allowed Caltex a 50 cent per barrel incentive payment on new and secondary recovery of crude oil production after January 1, 1977, thereby increasing Caltex's margin on secondary recovery production from an estimated $1.35–$1.40 to $1.85–$1.90 per barrel.[8] The production-sharing contracts signed in 1977 with several petroleum companies, including Elf Aquitaine, Mobil, Natomas, and Shell, also provided financial incentives, including recovery of all exploration and development investments in seven years instead of fourteen; a 20 percent investment credit for recovery of tangible costs associated with new petroleum exploration and development expenditures; and world market prices for the share of output each company must contribute to the domestic market for the first five years of oil production instead of a price equal to cost plus 20 cents per barrel.[9]

It is expected that the 1963 contracts of work held by Caltex and Stanvac will be renegotiated in the form of production-sharing contracts when they expire in 1983, and possibly Stanvac's contract of work that does not expire until 1993 will also be renegotiated in 1983.[10] Although the financial returns from the contracts of work (aside from the absence of bonus payments) do not differ substantially from those of the production-sharing contracts, the contracts of work have certain advantages. In the latter, management *legally* rests with the contractor, while under the production-sharing contracts, management *legally* is in the hands of Pertamina. A major

consequence is that purchases of equipment under the contracts of work are not subject to certain government approvals required under the production-sharing contracts.

1977 Model Production-Sharing Contract

In 1977 Pertamina issued a model production-sharing contract (to which minor amendments were made in 1980) that has constituted the basic format for contracts negotiated between Pertamina and private petroleum companies at least through 1981. Actual contracts differ with respect to signature and production bonuses, minimum expenditures on exploration, and certain other matters. Rather than reproduce the cumbersome text of the contract, the following is a summary of the major provisions of an actual agreement between Pertamina and a U.S. petroleum company (the name is confidential) negotiated in 1978. The principal provisions are believed to be typical of the production-sharing contracts negotiated during the 1977–81 period.

1. *Scope.* Pertamina has responsibility for the management of the operations under the contract, and the contractor is responsible to Pertamina for the execution of such operations in accordance with the provisions of the contract. The contractor provides all the financial and technical assistance required for such operations and bears the risks in carrying out operations.

2. *Term.* The term of the contract is thirty years from the effective date. If at the end of the initial six years no petroleum is discovered in the contract area, the contractor has the option of either terminating the contract or requesting two additional periods of two years each, which extensions shall be promptly granted. If after the initial six years and any extensions no petroleum is discovered in the contract area, the contract shall automatically terminate. If petroleum is discovered in any portion of the contract area within the period indicated above, which in the judgment of Pertamina and the contractor can be produced commercially, development will

[7]See "Aftermath of the Pertamina Debacle," *Petroleum Economist* (April 1977) p. 143.

[8]Office of International Affairs, *Role of Foreign Governments in the Energy Industries* (Washington, D.C., U.S. Department of Energy, October 1977) p. 364.

[9]Office of International Affairs, *Role of Foreign Governments*, p. 365.

[10]Caltex and Stanvac also have production-sharing agreements negotiated since 1963.

begin on that particular portion of the contract area. In other portions of the contract area exploration may continue concurrently.

3. *Surrender of Areas.* On or before the end of the initial three-year period from the effective date of the contract, the contractor shall surrender 25 percent of the original contract area; on or before the end of the sixth contract year (plus any extension) the contractor shall surrender an additional area equal to 25 percent of the original contract area; on or before the end of the eighth contract year the contractor shall surrender an additional area, so that the area retained shall not be in excess of 7,500 sq km, or 40 percent of the original total contract area, whichever is less.

4. *Expenditures.* Operations are to begin not later than six months after the effective date of the contract. The amounts spent by the contractor are to be not less than the following in each of the first five years:

First year	US$3.7 million
Second year	US$4.7 million
Third year	US$1.9 million
Fourth year	US$3.3 million
Fifth year	US$1.9 million

At least one exploratory well must be drilled during the first year of operation. Amounts spent in excess of the minimum in any one year may be carried forward to the following year, and, with Pertamina's consent, if the amount spent is less than the required minimum in any one year, it may be made up in the following year.

5. *Work Program and Budget.* Prior to the beginning of each year, the contractor must prepare and submit for approval to Pertamina a work program and budget of operating costs and operations for the ensuing year. Pertamina may propose a revision in the work program, but approval of the proposed work program and budget will not be unreasonably withheld.

6. *Compensation and Production Bonuses.* The contractor must pay to Pertamina as compensation for information $5.1 million within 30 days after the approval of the contract by the government and delivery of information. The

contractor must pay Pertamina $5.0 million after production from the contract area averages 25,000 bpd for 120 consecutive days; and $10.0 million after daily production from the contract area averages 75,000 bpd for 120 consecutive days. The contractor may not include these payments in operating costs for purposes of calculating output shares (paragraph 9), but they may be deducted in calculating taxable income.

7. *Rights and Obligations of the Contractor.* The contractor shall (a) advance all funds necessary for the execution of the work program; (b) have the right to sell, assign, transfer, or convey all of its rights and interests under the contract to any affiliated company or to a non-affiliated company with prior consent of Pertamina, which consent shall not be unreasonably withheld; (c) submit to Pertamina copies of all geological, geophysical, drilling, production, and other data and reports; (d) prepare and carry out programs for industrial training and education of Indonesians for all job classifications; (e) have the right to export its share of crude oil and retain abroad the proceeds from them; (f) give preference to Indonesian goods and services offered at equally advantageous conditions of price, quality, and availability; (g) pay the Indonesian corporate tax on net income as defined and the tax on interest and dividends; and (h) after commercial production begins, fulfill its obligation to supply the domestic market for petroleum in Indonesia.

8. *Calculation of Amount to be Supplied to Domestic Market.* The contractor agrees to sell to Pertamina a portion of its share of crude oil to be calculated as follows: The total quantity of crude oil produced from the contract area, times the ratio of that quantity to the entire Indonesian production of all petroleum companies, times 34.09 percent; or 25 percent of the crude oil produced from the contract area times 34.09 percent—whichever is the smaller amount. The latter amount represents 8.52 percent of the output after operating costs. Such crude is to be sold to Pertamina at 20 U.S. cents per barrel at the f.o.b. point of export. However, for a period of five years, starting with the first delivery of crude oil produced, the contractor will receive

the export price for the oil delivered to the domestic market.

9. *Share of Output.* The contractor may recover all operating costs (as defined in paragraph 21) out of sales proceeds. Of the crude remaining after deducting operating costs, Pertamina is entitled to receive 65.9 percent and the contractor 34.1 percent. Operating costs consist of (a) current-year noncapital costs; (b) current-year depreciation for capital costs; and (c) current-year allowed recovery of prior years' unrecovered operating costs.

10. *Marketing.* Normally the contractor will market all the crude from the contract area. However, Pertamina may elect to take its portion of crude oil in kind and export it if it can obtain a higher price.

11. *Investment Credit.* The contractor may recover an investment credit amounting to 20 percent of the capital investment cost required for developing crude production facilities. The investment credit may be recovered out of gross production in the earliest production year before tax deduction provided that, for the development project concerned, the quantity of crude oil to which Pertamina is entitled, together with 56 percent of the quantity to which the contractor is entitled, represent not less than 49 percent of the cumulative production over the project's life. This incentive may also be applied to new secondary recovery projects, but is not applicable to "interim production schemes" or further investments to enhance production within the primary production phase.

12. *Natural Gas.* If gas is produced and marketed, after deducting operating costs associated with natural gas production Pertamina is entitled to 31.8 percent and the contractor to 68.2 percent of the output.

13. *Valuation of Crude.* Crude is to be valued at the net realized f.o.b. price Indonesia for crude sold to third parties. Crude sold to other than third parties is to be valued by using the weighted average per unit price received by the contractor and by Pertamina from sales to third parties. If Pertamina can obtain a higher price than the contractor for the oil representing the operating costs, the contractor must either match

the Pertamina price or permit Pertamina to market the oil. In the latter case, Pertamina reimburses the contractor on the basis of the higher price. Contractors are prohibited from giving a discount or paying commissions to their affiliates.

14. *Title to Equipment.* Equipment purchased by the contractor becomes the property of Pertamina when landed at an Indonesian port. This does not apply to leased equipment belonging to third parties who perform services as a contractor.

15. *Arbitration.* If disputes arising between Pertamina and the contractor cannot be settled amicably, each party will appoint one arbitrator, and the two arbitrators will appoint a third. If the parties do not agree on a third arbitrator, he shall be appointed by the president of the International Chamber of Commerce. The arbitration shall be conducted in accordance with the rules of arbitration of the International Chamber of Commerce.

16. *Employment and Training of Indonesian Personnel.* The contractor agrees to employ qualified Indonesian personnel in its operations, and after commercial production begins will undertake the training of Indonesian personnel for labor and staff positions, including executive management positions. Costs of training are to be included in operating costs.

17. *Termination.* At any time following the end of the second year of the contract, the contractor may, after consultation with Pertamina, relinquish its rights and be relieved of its obligations under the contract.

18. *Processing of Products.* The contractor agrees to refine in Indonesia 10 percent of the contractor's share of the crude. Should no refining capacity be available, the contractor will establish refining capacity provided (a) Pertamina requests the contractor to do so; (b) the contractor's share of crude oil is not less than 100,000 bpd; and (c) the erection and use of such refining capacity is judged to be economical by both parties.

19. *Participation.* Pertamina has the right to demand that a 10 percent interest in the total rights and obligations under the contract be

offered to either a company designated by Pertamina, the shareholders of which shall be Indonesian nationals, or to an Indonesian entity designated by Pertamina. This right will lapse unless exercised by Pertamina no later than three months after the contractor's notification of its first discovery of petroleum in the contract area. Payment for the 10 percent interest to the contractor will be an amount equal to 10 percent of the sum of the operating costs which the contractor has incurred up to the date of the discovery; 10 percent of the compensation paid to Pertamina for information; and 10 percent of the production bonus payments. The Indonesian participant may, at his option, pay for the interest in cash within three months of the date of its acceptance, or out of 50 percent of the Indonesian participant's production entitlements. In the latter case, the total payment is to be 150 percent of the amount as calculated above.

20. *Taxes.* The contractor is to pay the Indonesian income and dividend tax. Pertamina assumes other Indonesian taxes of the contractor, including the transfer tax, import and export duties on materials, equipment and supplies brought into Indonesia by the contractor and its contractors and subcontractors, and other taxes levied in connection with operations performed.

21. *Definition of Operating Costs.* Operating costs consist of (a) current year's noncapital costs; (b) current year's depreciation for capital costs; and (c) current year's allowed recovery of prior year's unrecovered operating costs. Noncapital costs, including exploration and intangible drilling costs, may be expensed, and capital costs are depreciated by the double declining balance method. Depreciation schedules for capital expenditures made for items which have a useful life beyond the year incurred are stated in the contract. Interest on loans obtained from affiliates or parent companies or from third party nonaffiliates at rates not exceeding prevailing commercial rates for capital investment in petroleum operations may be recovered as operating costs.[11]

Joint-Venture Form of Production-Sharing Contracts

Beginning in 1977 Pertamina began negotiating joint-venture contracts with foreign petroleum companies in which each party contributes 50 percent of all expenditures for both exploration and production and share in the output on a 50-50 basis. However, with respect to the contractor's share, Pertamina receives 65.9 percent and the contractor 34.1 percent of the crude output after deducting the contractor's share of operating costs. Except for the joint-venture feature, the contract follows the same format as the production-sharing contract described above.

The first joint-venture contract was signed with Conoco in October 1977 for an onshore block in Irian Jaya. Conoco agreed to spend $15 million on exploration during the first three years. The signature bonus was $3.15 million and production bonuses were $1 million for over 50,000 bpd and $1 million more for production over 150,000 bpd. If oil is discovered, Conoco agrees to sell 5 percent of its interest to Indonesian enterprises.[12] Several joint-venture contracts were signed in 1978 and 1979, including an offshore contract between Pertamina and Mobil in March 1979. However, in 1980 and 1981 most of the contracts negotiated were regular production-sharing agreements.

It is reported that Pertamina proposes a joint-venture arrangement only when it believes the contract area is well explored and not very risky. By sharing in the initial exploration and development costs, Pertamina tends to gain substantially more revenue. One of the reasons given by the Indonesian Mining Minister for using the joint-venture arrangement is that "Pertamina personnel will be working directly with more experienced foreign personnel." The petroleum operations are to be conducted by an operator in accordance with policies, programs, and budgets approved by an operating committee. However, the contract states that the contractor shall be the operator for the duration of the contract, as long as it has a participating interest.

---

[11] In the case of this particular contract, the contractor agrees not to finance operations with loans.

[12] "Indonesia," *Petroleum News* (January 1981) p. 22.

Petroleum Loan Agreement

In 1980 Pertamina introduced a new type of petroleum agreement called the "petroleum loan." According to an agreement between Pertamina and the Indonesian Nippon Oil Company (Inoco), the latter made a loan of $16.0 million to Pertamina for the exploration and development of three oil fields, with the loan to be repaid by a share in production. Inoco is also entitled to purchase 40 percent of the output at the market price prevailing at the time of shipment for a period of ten years after the initiation of commercial production.[13] If no oil is found, Inoco receives nothing. Inoco has no operating responsibilities under the loan agreement.

The Contractor's Share Under Production-Sharing Contracts

As noted above, the contractor receives 34.9 percent of output after cost recovery, but since he pays a 56 percent income and dividend tax, he is left with roughly 15.4 percent of the output. However, this does not take account of the large signature and production bonuses which are not deductible in calculating operating costs. For example, the Gulf Oil production-sharing agreement of 1979 provided for a signature bonus of $20.1 million and production bonuses of $10 million for output in excess of 50,000 bpd; $50 million for output in excess of 100,000 bpd; and $100 million for output in excess of 200,000 bpd.[14] In addition, after the first five years the contractor must supply up to 25 percent of his share of the output after cost recovery to the domestic market at 20 U.S. cents per barrel. Finally, the contractor must offer 10 percent of his equity in the project to an Indonesian entity at cost. All of these factors greatly complicate the problem of estimating the contractor's share of revenue after accounting for all costs, and one would have to know the output over time in order to determine the effects of the bonuses

as well as other aspects of the contract in order to determine the contractor's share.[15] The contractor's share of the revenue after cost is sensitive to the price of oil and to the quality of the field. On the basis of 1982 prices and taking into account the new 20 percent investment allowance, Kemp and Rose calculate the contractor's share of net revenue in constant prices to range from 18.2 percent for a low-cost/high-volume field to 10.7 percent for a high-cost/low-volume field, and the corresponding real IRRs from 109.9 percent to 6.5 percent.[16] However, the actual split would be affected by the amount of the bonus payments, which differ considerably among the contracts.

Most Indonesian oil fields are small and the Indonesian fiscal system does not provide a satisfactory rate of return on small fields. However, the present PSCs that provide for full recovery of costs, as contrasted with the pre-1977 PSCs that limited cost recovery to 40 percent of gross revenue, have resulted in a high proportion of gross revenues going to the contractors producing small fields. In July 1981 Pertamina ordered Arco to cease production from its Sembakung field, asserting that all the oil produced from the concession had gone to Arco to pay development expenses, with none available as Indonesia's share. A compromise was reached in the dispute in July 1982 whereby Arco was allowed to recover costs before relinquishing the concession. *Petroleum News* suggests that contractors may find it difficult to obtain permission from the Indonesian government to develop marginal fields in the future because Pertamina does not want to relinquish a high share of the output to cover costs.[17] It seems likely that under the pre-1977 contracts that limited cost recovery to 40 percent, contractors would have been unwilling to develop small or marginal fields in Indonesia; but

[13]"Exploration Annual 1981," *Petroleum News* (January 1981) p. 22.
[14]*Petroleum News* (January 1981) p. 26.

[15]In discussing this question with company officials in Jakarta, I was told that the contractor's share of output after deducting bonuses and operating costs is probably less than 10 percent.
[16]Alexander Kemp and David Rose, "Four Oil Fiscal Systems in Need of an Overhaul," *Financial Times Energy Economist* (April 1982).
[17]"Indonesia," *Petroleum News* (January 1983) p. 17.

evidently Pertamina is also reluctant to pay the cost of developing small fields.

One of the reasons given for the high level of interest on the part of petroleum companies in Indonesian contracts and the record number of new PSCs signed by Pertamina during 1982 is the "total cost recovery" provision of the contracts. It appears, however, that where Pertamina believes the total cost is too high, it will not sanction field development.[18]

Competition for Production-Sharing Contracts

Pertamina is reportedly offering new areas for competitive bidding as rapidly as possible within the limits of its administrative capability and available personnel, including geologists. In 1980 and 1981, competition for contracts was reported to be keen among petroleum companies, and more than a dozen PSCs were signed in 1982 despite the decline in oil prices.[19] The winning bid is decided by Pertamina on the basis of the signature bonus; the amount the bidder proposes to spend during the first six years/of the exploration program; the production bonus; any offer of government participation in the venture in excess of the minimum requirement of 10 percent; and other considerations, such as a commitment by the company to build a refinery. Bonus payments are deductible for calculating taxable income, but are not recoverable as operating expenses. Integrated petroleum companies are often willing to accept a relatively low IRR on operations in order to be assured of oil supplies for their downstream affiliates; this provides a competitive advantage over nonintegrated companies in bidding.

Production Outlook

As of the end of 1982, Pertamina had signed over 80 contracts of work, production-sharing contracts, and joint-venture contracts involving individual petroleum companies or groups of companies representing a large portion of the major international petroleum companies of the world. In a number of cases, the same companies have negotiated more than one contract.[20] A number of contracts have been cancelled, but new contracts are signed every year, with contracts for eleven blocks being signed in 1980, nine in 1981, and thirteen in 1982. For the contracts signed in 1981, ten foreign companies gave Pertamina $100 million in signature bonuses and agreed to spend at least $560 million. Total spending for exploration and development during 1981 has been estimated by *Petroleum News* at just over $1.0 billion, an increase of about $150 million over 1980.[21]

Indonesia's petroleum production peaked at nearly 1.7 million bpd in 1977 and declined to 1.1 million bpd in 1981, rising again to an estimated 1.3 million bpd in 1982. Estimated proved reserves were less than 9.6 billion barrels at the end of 1982, down from 9.8 billion barrels in 1981. A report by Joseph P. Riva, Jr. (Library of Congress) concludes that "Much of Indonesia's remaining oil will be more difficult to find and produce, as it is likely to occur in smaller fields or in more difficult environments, such as offshore in deeper waters or in remote jungles. To find and produce even a modest portion of this remaining oil will require a much more intensive and expensive effort than has been expended to date, as more than half the oil which has already been discovered existed in 13 giant and near-giant fields that were relatively easy to find and produce. Any new giant fields will most probably be found in little-explored areas rather than in the areas which have been drilled for many decades. Smaller fields will continue to be found, but it will take

[18]Ibid., p. 23.
[19]"Asia/Pacific E&P Activity Holds High." *Oil and Gas Journal* (May 16, 1983), pp. 55–56.
[20]A complete list of these contracts, together with pertinent information, is given in *Petroleum News* (January 1983) pp. 18–23. Over thirty companies are involved, including Agip, Amarada Hess, Atlantic Richfield, British Petroleum, Caltex, Chevron, Cities Service, Conoco, Dominex, Exxon, Getty, Gulf, Houston Oil, Husky Oil, Huffco, Mobil, Mapco, Marathon Oil, Natomas, Philips Petroleum, Shell, St. Joe Petroleum, Standard Oil of California, Tesoro, Texaco, Total, and Union Oil.
[21]*Petroleum News* (January 1982) p. 24.

many of these to offset production declines in the older fields."[22] In contrast to oil, production of natural gas has increased since 1976 and Indonesia has recently initiated the production of liquified natural gas for export.

The reduced outlook for discovery of large fields and the increasing costs of discovery and production suggest that Indonesia should restructure its fiscal arrangements so as to make high-cost/low-volume fields financially attractive to petroleum companies.

Comments on the Revenue Provisions of the Indonesian Production-Sharing Contracts

The revenue provisions of the Indonesian production-sharing contracts involve several devices, so that the evaluation of their efficiency as revenue producers is complex. Except for the signature bonus payable at the time of signing of the contract, the bonuses are not front loading but are a form of nonrecoverable production costs that apply to both the contractor's and Pertamina's share of the output. Since the production bonuses rise as production increases, they could reduce the efficiency of the development of the contract area. However, Pertamina may seek to avoid efforts on the part of the contractor to keep production below levels that trigger higher bonus payments through its right to approve the annual work programs. In October 1981 Pertamina halted production by a joint venture involving Arco and Philips Petroleum in Borneo, reportedly on the grounds that the output of some 10,000 bpd was less than the potential output of the contract area.[23] Pertamina is said to have been dissatisfied with Arco's exploration effort. This was apparently the first time that Pertamina had taken an action of this kind.

Production sharing takes place after operating costs are deducted so that its effect is similar to a tax on net income, in contrast to that of a royalty. The obligation to deliver a portion of the contractor's crude to the domestic market at virtually no net return is also similar to a tax on net income.

Overall, the Indonesian system is more efficient from the standpoint of maximizing output than a system of straight royalties, but is probably less efficient than a system in which the entire revenue of the government comes from taxes on net income. As shown in table 4-4, the Indonesian system is regressive in that the government take as a percentage of net revenues rises with lower quality fields. Also, the IRRs for the high-cost/medium-volume and high-cost/low-volume fields are unacceptable according to the simulations by Kemp and Rose. On the other hand, the IRR for the medium-cost/medium-volume and low-cost/high-volume fields are quite favorable, which may help to explain the relatively high interest shown by private companies in Indonesian PSCs.

Discovery rates in Indonesian waters are relatively high and in 1981, 70 of 174 exploration wells drilled in the first nine months hit oil, gas, or a mixture of both—a success ratio of 45 percent. However, the discoveries tend to be small and, given the fact that most Indonesian oil fields are declining at a rate of 1.0 to 1.5 percent per year, reserves yielding an additional 200,000 bpd must be discovered each year in order to maintain production.[24] A recent assessment by a petroleum geologist is as follows.

Given the advanced age of some of the larger fields in the country and the maturity of oil exploration in many of the basins, it would appear that a major effort to maintain oil production in the older, large fields by secondary and perhaps even tertiary methods where possible will have to be coupled with the discovery of new fields of significant size to result in any production increases that could be sustained over the next several years. This is not impossible, but it may not prove likely.[25]

[22] "Petroleum Prospects of Indonesia," *Oil and Gas Journal* (March 8, 1982) pp. 315–316.
[23] "Indonesia Halts ARCO Oil Output at Site in Borneo," *Wall Street Journal* (October 16, 1981) p. 4.

[24] *Petroleum News* (January 1983) p. 26.
[25] J. P. Riva, Jr., "Petroleum Prospects of Indonesia," *Oil and Gas Journal* (March 8, 1982) pp. 306–316.

6

Peru's Production-Sharing Contracts

Peru initiated the PSC format in 1971 following a long period during which no new private petroleum contracts were negotiated. As in the case of the Indonesian PSCs, Peru's model contracts have gone through substantial changes, in part as a consequence of the change in U.S. tax regulations in 1977, and in part in response to efforts on the part of the Peruvian government to extract an increasing share of the economic rents. However, Peru's petroleum potential is far less than that of Indonesia (with estimated proved reserves of less than 10 percent of Indonesia's). Exploration risk is quite high, and Peru has attracted a relatively small number of companies to negotiate contracts in recent years.

Historical Background

Commercial production of crude petroleum in Peru dates from the nineteenth century when the Talara field in northwestern Peru was first drilled as early as 1863. Prior to the military coup of October 1968, Peru's petroleum was produced by private firms under concessions and by the government enterprise, Empresa Petrolera Fiscal (EPF). In 1966 Peru produced 23.0 million barrels of crude, of which about 75 percent was accounted for by the International Petroleum Company (a subsidiary of Standard Oil of New Jersey); about 10 percent by EPF; about 7 percent from offshore operations by Belco Petroleum Corp.; and the remainder by several small private operations. At that time, 93 percent of the 41 million hectares under concession were held by EPF and the remaining 2.8 million hectares by 15 private foreign and domestic companies. No new private exploration concessions were granted after July 1956, although some exploration concessions were converted to exploitation concessions after that time. The International Petroleum Corporation was expropriated by the military government in 1968.

The new military government shifted from the concession system to one in which exploration/exploitation contracts were negotiated with private companies by the government petroleum agency, Petroleos del Peru (Petroperu)—formerly EPF. In 1971 Petroperu signed its first contract with Occidental Petroleum for blocks in northeastern Peru. This contract followed the

Indonesian pattern of production sharing, but incorporated certain other features to form what has been called the 1971 "Peruvian Model." Two contracts with Tenneco/Union Oil and one with British Petroleum were also negotiated in 1971 and followed the same pattern. Subsequently, virtually all private concessions were relinquished (either voluntarily or involuntarily) and Belco Petroleum's offshore concession was converted into a Peruvian model contract with Petroperu in 1973.

By the end of 1973, Petroperu had signed contracts with 18 private companies or consortia, 16 of which were for exploration/exploitation in the eastern jungle, and two for offshore operations. Only Occidental found commercially marketable oil in the jungle (near the Ecuadorian border) and by the end of 1976 all foreign contractors except Occidental and Belco had left the country after unsuccessful drilling results (including finding oil that was not commercially marketable). There is also some evidence that the companies were discouraged by the general political climate in Peru.

The government suspended further negotiation of contracts in September 1973, and no new contract areas were available until the publication of the conditions governing new contracts in March 1977. There is no rational explanation for this suspension since there were 36 million hectares of potentially oil-bearing areas undeveloped in 1973, and Peru's crude production in 1973 (69.0 million barrels) was below the 1970 level (72.2 million barrels). Most observers in Peru attribute the suspension to a whim of the then Minister of Energy and Mines, General Jorge Fernandez Maldonado. Maldonado was a left-wing nationalist and was probably prejudiced against multinational petroleum companies.

The continuing decline in crude production despite substantial investment by Petroperu led the government to seek new contracts with private firms in 1977, but no contracts were signed that year mainly because of a lack of interest on the part of private companies. In April 1978 two contracts were signed with Occidental: one for a heavy oil area in the eastern jungle (block 1B) and another for secondary recovery in the Talara area.

The 1971 Model Contract

According to the 1971 model contract between Petroperu and Occidental (and subsequently with 17 other companies), the company provides all technical and financial resources for carrying out exploration and development (except for building the Trans-Andean pipeline). The contracts were for thirty-five years from the date of signing, and all petroleum produced was to be divided between Petroperu and the private contractor at the wellhead. Under the Occidental contract for exploration/exploitation in the northwestern jungle area, 50 percent of total output produced went to Petroperu and the remainder could be exported by Occidental or sold at the world market price to Petroperu, subject to the need to fulfill Peru's domestic requirements. The split in oil production differed from contract to contract, ranging from 44 to 50 percent for the company, depending upon the assessment of risk, cost of development, and the volume of production. Unlike the early Indonesian PSCs, no provision was made for cost oil allowance before the production split.

Under the 1971 model contract, the companies agreed to undertake specified exploration programs during the first four years, which included seismic studies, photogeological surveys of the contract areas, and drilling of several wells; usually a minimum of four exploratory wells were required even if the prospecting results were unfavorable. The exploration programs were backed by a bank guarantee by the companies, and failure to undertake the agreed amount of drilling meant forfeiting the guarantee funds. Additional drilling commitments—one every five months—were provided from the 54th through the 79th month. The companies had seven years from the date of signing the contract to declare they had a commercial discovery that they planned to exploit.

Peruvian profits and remittance taxes on the companies were, according to the contract, to be paid by Petroperu, together with customs duties on imported equipment. However, the companies paid an equity share tax, a business license tax, taxes on real estate, stamp taxes, and certain labor law taxes and contributions; these taxes did not constitute a significant portion of gross income. The applicable income tax plus the 30 percent branch profits (remittance) tax added up to an effective rate of 47.5 percent for the first ten years; 54.5 percent during the subsequent ten years; and 65.0 percent thereafter for the duration of the contract. The tax was regarded as a heavy burden on Petroperu since it was paid out of Petroperu's 50 percent share of the output.

Remittances by the contractors were limited to net profits (as defined by law) plus depreciation. The Peruvian Central Bank guaranteed the availability of foreign currency for allowable remittance of net profits, depreciation, justified services, amortization of loans, and interest payable on loans.

The April 1978 Peruvian Model Contract

There were certain difficulties with the 1971 Peruvian model contract which were corrected by the model represented by an Occidental Petroleum contract dated April 1978 and based on principles announced by Petroperu in March 1977. A major difficulty with the old contracts was that the contractor was required to drill a minimum number of wells, even in cases where seismic and other data indicated that oil was unlikely to be found. The result was that several companies simply forfeited the bank guarantees without drilling the wells.

The Occidental contract of April 1978 for operations in the jungle area (block 1B) provided for two phases for the exploration period. Phase 1, which was to be completed within two years, required the contractor at his discretion either (a) to complete and production test a previously drilled exploratory well; or (b) to drill a new well. The contractor was also required during the first phase to install pipelines and production facilities necessary to conduct a pilot test of one of the wells and to conduct seismic and other geological and geophysical studies of

the contract area. Following the signing of the contract, the contractor was required to provide Petroperu with a bank guarantee of $8 million. If the contractor decided to relinquish the contract area before the completion of phase I, the entire $8 million was forfeited. However, on completion of phase I the penalty was reduced to $3 million. Phase II of the exploration period was to be completed within four years from the effective date of the contract; it included the drilling of two additional exploratory wells in the contract area. Upon completion of each exploratory well, the penalty was reduced by $1.5 million. If the contractor decided to relinquish the contract after drilling two additional wells, there was no further penalty.

The drilling of additional wells must begin no later than the end of the 48th month after the effective date; operations for drilling the second additional well must begin no later than the end of the 54th month; operations for drilling the third additional well must begin no later than the end of the 60th month; and operations for drilling the fourth additional well must begin no later than the end of the 66th month.

Within 30 days after the contractor declares a commercial discovery, he must deliver to Petroperu a program for the development of the contract area, and within six months the contractor must begin the execution of the program, which is subject to adjustment. The term of the contract is 30 years from the effective date.

Because the jungle area in block 1B contains heavy oil and there are special difficulties in its exploitation, Occidental received 75 percent of the heavy oil produced for the first 12 million barrels; 70 percent for the next 8 million barrels; and 60 percent for the next 10 million barrels. After 30 million barrels of heavy oil had been produced, the contractor would receive 50 percent of the output, with the other 50 percent going to Petroperu.

The tax arrangements under the 1978 model contract were similar to those under the 1971 model. All profits and remittance taxes and import and export duties were paid by Petroperu, leaving only minor taxes, such as the business equity share tax, tax on undeveloped land, and social security taxes, to be paid by the contractor. The contractor was permitted to remit revenues from the sale of his share of the output less domestic expenses free of profits and remittance taxes. Such remittances included amortization and interest payments on external debt of the contractor and payments for foreign services and supplies. Foreign exchange availability for such remittances was guaranteed by the Central Bank whether the petroleum was exported by the contractor or sold to Petroperu.

A second contract between Petroperu and Occidental in a joint venture with Bridas Exploration and Production Company (an Argentine firm) was also negotiated in April 1978 for secondary recovery operations in the producing Talara area of northwestern Peru. This contract provided that 49 percent of the output would go to the joint venture (84 percent of which was held by Occidental) and the remaining 51 percent to Petroperu. Otherwise, the contract is similar to the April 1978 contract with Occidental for operations in the jungle. From the beginning of its operations to early 1979, Occidental had invested $210 million in Peru.

No other contracts employing the 1978 model were completed, but Petroperu and Belco Petroleum had reached an understanding on a new offshore contract according to which 47.5 percent of the output would go to Belco and the remainder to Petroperu. However, completion of this contract was deferred pending the enactment of a new hydrocarbon law which is discussed below.

The Tax Issue

As has been noted, according to both the 1971 and the 1978 model contracts, the income and branch profits (remittance) taxes were to be paid by Petroperu out of its share of the output, so that private contractors had no effective tax liability to the Peruvian government. Until the end of 1977, U.S. petroleum companies were able to credit their tax liabilities to the Peruvian government (paid by Petroperu) against their U.S. corporate tax liabilities. Thereafter, U.S. firms were permitted to credit taxes paid to for-

eign governments against their U.S. corporate tax obligations only if the foreign tax systems met the standards established by Rev. Rul. 76-215 issued by the Internal Revenue Service (IRS). The 1976 Tax Reform Act deferred application of Rev. Rul. 76-215 until December 1977, and in the interim taxpayers were expected to renegotiate their contracts with foreign governments to conform with the IRS rulings. (See chapter 11 for a review of U.S. government regulations relating to foreign tax credits.)

In general terms the IRS required that in order to be credited against U.S. tax obligations, taxes paid to foreign governments had to be paid directly by the U.S. investor to the foreign government; must be calculated independently from amounts paid to the foreign government as owner of the minerals; and must constitute a tax on net income after allowance for expenses. Both U.S. petroleum companies and Petroperu were anxious to change the Petroperu law relating to petroleum contracts so that Petroperu was not liable for payment of the contractor's Peruvian income taxes and that the taxes paid by the contractor directly to the Peruvian government would meet the requirements of the IRS for crediting against U.S. tax obligations. Negotiations led by Belco were conducted with Petroperu and the Peruvian government for a new Peruvian petroleum tax system that would be satisfactory to all parties. New Peruvian petroleum contract and tax legislation was drafted in 1978, which apparently had the acceptance of all the Peruvian authorities involved. But for this legislation to be meaningful, it required a ruling by the U.S. Department of the Treasury that Peruvian tax payments made by U.S. petroleum companies would in fact be creditable against U.S. tax liabilities. The proposed Peruvian legislation provided for a tax of 40 percent on gross income of petroleum companies to be paid directly by the companies; in addition, a portion of the output would be retained by Petroperu, the amount being determined so as to arrive at the same split between the contractor and the government of Peru that was negotiated under the original contract. A favorable IRS ruling was not issued until the fall of 1979, following which it was expected that the new Peruvian petroleum contract and tax law would be approved and thereafter the petroleum contracts held by Belco and Occidental would be renegotiated in accordance with the terms of the new law.[1]

Meanwhile, there developed opposition to the proposed new petroleum tax law in the Peruvian Ministry of Finance and the law was never enacted. Instead, new decree laws were issued in December 1979 and January 1980 (DL 22774, 22775 and 22862 and Supreme Decree 010-80-EF) which in effect required a renegotiation of the original contracts under terms that were much less favorable to both Belco and Occidental in comparison with the terms of their existing contracts. Under the new decrees, the contractors were required to pay a tax on net income equal to the current standard tax applicable to foreign companies (68.5 percent) and to renegotiate the split in output between Petroperu and the contractor, with the contractor assuming the total cost of production for the entire output. This action by the Peruvian government was regarded by both Belco and Occidental as a breach of contract and a forced renegotiation which resulted in substantially less favorable terms. Renegotiations were completed in April 1980 and new contracts signed in July 1980 covering three Occidental contracts and the Belco contract.[2] Although the terms of the new contracts are similar, the summary and analysis in the following paragraphs apply also to the Occidental contract.

Contract Between Petroperu and Occidental Petroleum of July 1980

This contract constitutes a substantial revision of Occidental's contract for block 1A-A in northeastern Peru signed on June 22, 1971. The

[1] The interest of the Peruvian government was indicated by the fact that Peruvian officials visited the U.S. Department of the Treasury to request a favorable IRS ruling. The Treasury ruling of August 1979 provided that a foreign tax on gross income in lieu of a tax on net income would, under certain specified conditions, be creditable against U.S. tax liabilities.
[2] The revised Occidental contracts include the two negotiated in 1978 and one negotiated in 1971.

major provisions are given in the following paragraphs:

1. The contractor will be in charge of petroleum operations, but Petroperu will participate in the preparation of programs and their execution through a "supervisory committee" composed of representatives of the Peruvian government and of the contractor. The contractor must submit five-year development programs to Petroperu and to the supervisory committee.

2. The contractor is required to maintain a training program for Peruvian personnel at every level, aimed at the replacement of expatriates with nationals. The contractor is also obligated to avoid environmental pollution resulting from petroleum operations in the contract area.

3. The contractor is obligated to pay Petroperu on a monthly basis a bonus of 1.9 billion soles per year (approximately $5.6 million as of December 1980) for twenty years. In the event the contract is terminated before December 15, 2000, the contractor will immediately pay all remaining installments.

4. The contractor's share of the output is as follows:

Up to 150,000 bpd	50 percent
150,000 to 200,000 bpd	48 percent
200,000 to 250,000 bpd	46 percent
250,000 to 300,000 bpd	44 percent
In excess of 300,000 bpd	42 percent

5. The contractor is to receive 50 percent of the total volume of natural gas produced from the contract area and sold commercially, regardless of the volume produced.

6. Within the contract area, the contractor must bear all costs of production, storage, and transportation of petroleum to the principal pipeline.

7. Petroperu has the right to purchase an amount of the contractor's portion of petroleum output for internal consumption, the amount being determined by the ratio of Peru's total consumption to Peru's production. In addition, Petroperu is to receive an amount of petroleum equivalent to the

contractor's tax obligations to the government of Peru. The sales price of crude in the internal market is US$26 per barrel, adjusted cent-by-cent according to the average price fluctuations of a basket of international crudes.

8. Subject to the requirements of paragraph 7 above, the contractor is entitled to export the remainder of his share of the output. However, the contractor is obligated to pay Petroperu 50 percent of any amount by which the contractor's f.o.b. export price exceeds a base f.o.b. price. The base f.o.b. price (BP) at the date of the export sale is determined by the following formula:

$$BP = US\$36.00 + (CRP - US\$29.91)$$

where CRP is the average official sales price per barrel on the date of the export sale of (a) light Arabian crude f.o.b. Ras Tanura; (b) Qatar Marine f.o.b. Qatar; and (c) El-Sider, Libyan crude, f.o.b. El-Sider. Payments to Peru become mandatory under this formula when the actual sales price is $10 per barrel above the average of the official prices of the crudes listed above.[3]

9. The contractor pays a tax on net income from operations, but is protected from an increase in the Peruvian corporate income tax (68.5 percent) applicable to petroleum operations as of the date of the contract, and from any new taxes on petroleum operations. There is a formula according to which the contractor's share is raised with any increase in tax obligations. A similar formula applies to natural gas.

10. The contractor is responsible for paying any other taxes applicable to its operations in Peru in existence as of January 1, 1980 except that the following taxes will be for the exclusive account of Petroperu: (a) royalties on petroleum production; (b) taxes levied on the exportation of petroleum; and (c) customs duties on capital goods and raw materials imported by the contractor.

11. By a contract with the Peruvian central bank, the contractor is guaranteed the availability of foreign currency required to remit depreciation, amortization, and after-tax profits; foreign expenditures for commodities and services; and service on foreign indebtedness.

However, the contractor is obligated to deposit foreign currency receipts from export sales in the central bank's account at a foreign bank, against which the contractor will have the right to draw funds for authorized purposes.

12. The term of the contract is effective from January 1, 1980 to June 21, 2006 unless terminated earlier. Upon termination of the contract for whatever reasons, the contractor shall deliver to Petroperu free of charge, in working order or usable condition, any buildings, installations, pumps, machinery, and other assets constructed and used by the contractor in Peru which are related and/or accessory to the operations under the contract at the time of termination and are the property of the contractor.

The provisions of the renegotiated Belco agreement of July 1980 are essentially the same as those of the Occidental agreement summarized above except for a smaller annual payment and a somewhat different schedule in terms of output per day for the production split between the contractor and Petroperu. In September 1981 Belco signed a contract for another offshore area.

Although the 1980 renegotiated contracts with Occidental and Belco were in accordance with the U.S. tax provisions for crediting foreign income taxes, they were much less favorable to the companies than the original contracts since, in addition to the production split with no allowance for cost recovery, the companies are subject to the 68.5 percent corporate income tax. These conditions apparently reduced the companies' enthusiasm for expanding their investments and, in addition, delayed negotiations with other prospective contractors and generally reduced the interest of international petroleum companies in investing in Peru.

Contracts with Shell and Superior Oil

Royal Dutch Shell's Peruvian subsidiary, Shell Exploradora y Productora del Peru, began negotiating a contract for the exploration and development of a large area in southeastern Peru in 1979. Shell officials resisted the contract terms more or less imposed on Occidental in the rene-

---

[3]This occurred in the case of two sales early in 1980.

gotiated contract of July 1980 and were particularly averse to the bonus payments and high corporate income tax (68.5 percent). Except for the bonus payment, the contract that was finally signed on July 10, 1981 was similar in many respects to the July 30 contract between Petroperu and Occidental Petroleum. Under the contract, Shell is obligated to spend $100 million on exploration during the six-year exploration period.

Since the northern pipeline serving the area of Occidental's operations could not serve Shell's area in southeastern Peru, a new pipeline for transporting the oil to the coast would have to be constructed. Therefore, in the contract with Shell there were three schedules setting forth the production split, depending upon whether (a) the contractor would be wholly responsible for financing construction of the pipeline; (b) whether Petroperu decided to finance the pipeline construction entirely; or (c) whether the pipeline would be financed jointly by Shell and Petroperu. Without knowing the cost of the pipeline, it is difficult to compare the schedules of production split between Shell and the Occidental contracts. However, in the event that Shell is wholly responsible for financing the pipeline, the contractor's portion of the output is much more favorable, at least for outputs under 160,000 bpd. For production up to 90,000 bpd, the contractor's portion is 67 percent, declining to 60 percent for output between 120,000 and 160,000 bpd. However, in the event that Petroperu decides to finance the pipeline, the contractor's portion of the output up to 120,000 bpd is 50 percent, declining by steps to 25 percent for output in excess of 200,000 bpd.

During the course of negotiations with Shell, there were important changes in the Peruvian corporate income tax law applying to foreign petroleum companies. The most significant changes were embodied in a law dated December 26, 1980 that provided that all private and mixed companies operating north of the seventh parallel in the jungle and in the Talara area were eligible for a deduction of up to 40 percent of their pretax income for investments in petroleum and development; while companies operating south of the seventh parallel in the southern jungle and on the coast were eligible for a 50 percent reinvestment credit.[4] Over time, these reinvestment tax credits would serve to reduce substantially the effects of the high corporate income tax as field exploration and development progressed. According to a statement to the author by the chairman of Belco Petroleum, the investment credit will enable Belco to expand its exploration and development program in Peru.

In March 1981 the contract between a joint venture involving Superior Oil (90 percent) and Phoenix of Canada (10 percent) for exploration in northeastern Peru (near Occidental's contract zone) was signed. This contract is similar to the Shell contract described above except for guaranteed exploration expenditures, the percentages of output going to the contractor, and the arrangements relating to construction of the pipeline. (Oil produced in this area could presumably use the existing northern pipeline.) In this contract Superior Oil provides a bank guarantee of $10 million for exploration expenditures during the first two years and, if oil is found, agrees to make a total investment in excess of $20 million. The shares of output going to Superior Oil are the same as those provided in the Occidental contract of July 1980. The formulas for prices to be paid to the contractor for oil sold to the domestic market or for export are the same as those in the Shell and Occidental contracts, except for the base price, which reflected prices at the time the contract was signed.

A contract for a jungle block was signed by Hamilton Brothers in early 1982, and negotiations were reported for other jungle blocks with Cities Service, Union Texas, Marathon and Belco, but no agreements had been announced at the time of writing.

Comments on the Peruvian Petroleum Contract System

The Peruvian contract system described above is not well designed either to maximize Peru's

[4] "Peru: New Incentives for Private Sector," *Petroleum Economist* (February 1981) p. 79; see also, "Peru Adopts Incentives to Develop, Explore Its Petroleum Reserves," *Wall Street Journal* (December 30, 1980) p. 2.

income from petroleum revenues or to attract investment in exploration. It contains many regressive features and the overall government take is so high that the development of high-cost/low-volume fields is unlikely to be profitable. The bonus system, although providing for monthly installments, obligates the contractor to pay the full amount due over a twenty-year period even if the contract is terminated beforehand. In the case of the Occidental contract for block 1A-A of July 1980, the bonus is 1.9 billion soles annually for twenty years.[5] The bonus payments are, however, deductible in calculating taxable income, so that the effective tax is less than one-third of the amount of the bonus payment. Nevertheless, the obligation may be sufficient to deter some petroleum companies, especially since it must be paid even if the operation is not profitable. It should be noted that the Shell and Superior Oil contracts do not provide for bonus payments.

The production-sharing arrangement works against the efficient development of a field since, unlike the Indonesian PSCs, no allowance is provided to enable the contractor to recoup the

costs of Petroperu's portion of the output before the production split. Moreover, the schedule whereby higher levels of output increase Petroperu's share is a further deterrent to the exploration and development of marginal areas.

The income tax rate of 68.5 percent, without accelerated depreciation, provides only a small potential return for a high-risk country in which large fields and high-producing wells are unlikely to be found. However, the investment credit will reduce the effective rate during the period of investment following the initiation of production. Judging from the experience of more than a dozen companies that have explored in the eastern jungle of Peru without commercial discovery, the probability of a discovery appears to be less than 10 percent.

Let us assume an average production split of 48-52 in favor of Petroperu and that the total cost of producing 100 barrels of oil is 30 barrels, of which 10 barrels represents amortization of exploration and development costs, and 20 barrels is for operating expenditures. Without allowing for the reinvestment credit, this would leave the contractor with an after-tax net return of less than 5.7 barrels. Given the low probability of a commercial discovery, this appears to be a rather low return on a high-risk investment.

[5]No mention is made in the contract for the revaluation of the bonus payments with a depreciation of the sole in terms of the dollar. If no provision has been made for revaluation of the monthly or annual payments, such payments are likely to decline rapidly in dollar equivalent.

7

Guatemala's Production-Sharing Contracts

Guatemala's production-sharing contracts differ from the original Indonesian model in that they are awarded by competitive bidding for the contractor's share and for various other commitments in the exploration program. Income and withholding taxes are paid by the state as a part of its share of the output, as was the case in the Indonesian production-sharing agreements prior to 1977. The contractor is responsible for all exploration, development, and operating expenses. There is no provision for recovery of expenditures before the production split. Few companies have been attracted to bidding for these contracts, in part because little oil has been found in Guatemala and in part because of the deficiencies and general unattractiveness of the contracts. Increased political risk may also be a factor, especially since 1981.

Early Exploration: Background of Present Legislation

Various petroleum companies have been exploring in Guatemala for at least a quarter cen-

tury, but oil was not found in commercial quantities until 1974. Prior to July 1974, the government granted concessions for petroleum exploration and production. Concession holders were required to make a minimum investment each year and to pay a royalty of 12.5 percent of the value of oil produced. In addition, they were subject to the regular tax on net corporate income. A number of concessions were granted, but by 1974 only those of Petromaya and Centram (initially a joint venture of Inco and Hanna Mining) remained in force. However, after Petromaya discovered oil on its concession in central Guatemala, there were a large number of applications for concessions covering a substantial portion of the country. The government decided that the concession system would not provide a sufficient return to the government if a substantial amount of oil were produced and, in addition, it was concerned that the concession system might lead to a variety of abuses and be a source of corruption. Moreover, as in nearly all developing countries, private petroleum exploitation and international petroleum companies in particular are politically unpopular. The government wanted to devise a system of contracts which would assure that a substantial portion of the revenues would go to the government, and that the government would maintain control over the national petroleum resources.

Model Contract for Petroleum Exploration and Exploitation

The National Petroleum Law Decree 62-74 providing for a system of contracts became effective July 1, 1974, and, although new concessions were no longer granted, the existing concessions held by Petromaya and Centram remained valid until 1980. The 1974 law was found to contain defects and was replaced by Decree 96-75, which became the basis for the contracts that have been negotiated between the government of Guatemala and private petroleum companies. Decree 96-75 states that contracts with private companies will be signed by the minister of economy and approved by the president of the republic. Guatemala has no state petroleum enterprise and the government is not directly engaged in petroleum operations.

The following discussion of the model contract for petroleum operations is based on the following four documents:[1]

1. The Model Contract for Petroleum Operations of Exploration and Exploitation, January 11, 1978;
2. Regulations for the Exploration and Exploitation of Hydrocarbons, January 11, 1978;
3. Call for Submission of Bids and Awarding of Areas Designated for Exploration and Exploitation of Hydrocarbons, January 24, 1978; and
4. Selection of Areas Designed for Exploration and Exploitation of Hydrocarbons, January 11, 1978.

No attempt will be made to analyze or even provide a comprehensive summary of these

lengthy documents. The discussion will seek to highlight those aspects of the model contract and the method of bidding for the contracts that are of special importance for potential investors in Guatemalan petroleum exploration and development.

1. The contractor must submit and carry out within the contract area an exploration program approved by the government. During the exploration period the contractor is committed to drilling a minimum number of wells to a minimum depth and is required to guarantee a minimum amount of investment during each of the first six years of the contract by posting bonds. Although minimum requirements are established for each area before bidding for the contracts, the competitive bids may establish requirements in excess of the minimums.

2. A total minimum investment for the first three years of the exploration period must be guaranteed by the contractor before signing the contract. The contractor must also guarantee a minimum investment for exploration before the beginning of each annual period for the fourth, fifth, and sixth contractual years. However, if the contractor relinquishes the total exploration area before the beginning of any of the fourth, fifth, or sixth annual periods, the contractor will be exempt from his obligation to guarantee the minimum investments corresponding to the contractual years following the relinquishment.

3. Each time the contractor selects a block for inclusion in the exploitation area, he must submit an exploitation program for government approval. The contractor is committed to guarantee a minimum investment during each year of the exploitation period corresponding to the projection of his investment program.

4. All financial commitments are adjusted annually for inflation and for changes in the international value of the quetzal (Q).

5. There is a graduated minimum state participation in gross production specified for each of the contract areas, ranging from 55 percent for production up to 15,000 bpd and up to 75 percent for production exceeding 100,000 bpd.

These documents were published in the Guatemalan government's *Official Gazette* and English translations were available in the form of mimeographed documents.

Again, state participation may be higher if the contractor's competitive bid is beyond the minimum participation requirement. The government has the option to receive its participation in cash at a price determined by the international market price, or in oil. There are also minimum state participation rates for natural gas.

6. The state's participation in output includes the amount of income tax and any other tax on capital or income from petroleum exploitation. (Other taxes are discussed below.)

7. During the exploration phase the contractor is committed to specific expenditures each year for roads, and to the building of schools and hospitals during the production phase. The contractor is also committed to carrying out training and scholarship programs and to giving preference to qualified Guatemalan technicians.

8. The contractor is exempt from customs duties on imports required to carry out the terms of the contract.

9. The contractor is free to export or sell domestically his share of the hydrocarbons produced.

10. The duration of the contract is twenty-five years, including both the exploration and exploitation phases. The contractor must progressively renounce portions of the area so that at the end of five years he will have renounced 50 percent of the area assigned. Except in special cases, the area selected and retained for commercial production may be no larger than 10,000 hectares.

11. No arbitration procedures are provided in the contracts. Issues related to the interpretation and compliance of the contracts are subject to the jurisdiction of the Court of the Department of Guatemala and the contractor agrees not to use any form of diplomatic action.

Submission of Bids and Awarding of Areas to Contractors

From time to time the government solicits bids for specified areas subject to minimum requirements. In evaluating the bids for purposes of

awards, the following criteria are applied in an excluding order:

a. The number of wells offered to be drilled to at least the minimum depth.
b. The percentage participation of the state offered for the level of production of crude oil up to 50,000 bpd.
c. The percentage participation of the state offered in the production of crude in amounts over 50,000 bpd.
d. The depth of the wells to be drilled in excess of the required minimum.

Each bidder must pay an administrative fee for the submission of each bid equal to Q50,000.[2] In order to maintain the validity of the bid, a bond in favor of the state must be given equivalent to 5 percent of each of the following items: (a) 1 million quetzales which must be paid for the signing of the contract; (b) the amount of the minimum investment for the first three years; and (c) the amount of the commitments for training and other contract obligations for the first year.

Taxation

Despite the fact that the government's participation in production includes the income tax, the regulations for the administration of the contract make the contractor responsible for paying the taxes on net income in accordance with the income tax law directly to the internal revenue office, and for supplying all the necessary documents. The government will then reimburse the contractor for the total amount of the taxes within 15 days following the date the payment was made. However, the government will not reimburse the contractor for any amount in excess of the value of the state's participation for the corresponding fiscal period.

There is a graduated income tax with amounts in excess of Q500,000 subject to a 48 percent tax. In addition, profits remitted abroad or cred-

[2]The quetzal (Q) is at parity with the U.S. dollar.

ited to the account of a nonresident are subject to a maximum withholding tax of 11 percent. The special tax reimbursement system applicable to petroleum operations includes both the income tax and the withholding tax on profits remitted abroad.

Petroleum contractors are subject to the following taxes that are not covered by the tax reimbursement system.

a. An export tax of 2 percent based on the international market price of petroleum on the date of production.

b. A fiscal stamp tax of 2 percent payable on the face value of practically all accounting documents such as receipts, contracts, notes, etc. This could amount to a sizable burden on petroleum companies, but it is not clear as to just what amounts the 2 percent tax will be applied.

c. A contract subscription tax of Q1 million (bonus).

d. Social security taxes of 11.3 percent of payrolls and other remunerations to employees.

Production-Sharing Contract Awards, 1978–82

In January 1978 the Guatemalan government issued invitations for bids on six areas which were the subject of past requests by petroleum companies interested in exploration. All were onshore and wholly or partly in the northern province of Peten, and two were near the Mexican border (see figure 7-1). Only four companies or consortia submitted bids on four of the six areas offered. A Texaco-Amoco joint venture bid on two areas, blocks D and E, adjacent to the Mexican border, but were allowed only block D. A joint venture headed by Getty (and including Monsanto, 25 percent, and Texas East, 25 percent) and the French government enterprise, Elf Aquitaine, submitted competing bids for area BB in the central part of the country, but the consortium headed by Getty won the contract for the area. A consortium headed

by Hispanoil (Spain) and including Elf Aquitaine and Braspetro (Brazil) bid on area AA, adjacent to the area won by Getty, without competition. Since the Getty consortium had competition, it was required to bid more than the minimum well and investment requirements established by the government.

No other PSCs were awarded until December 1980 when the Hispanoil-Elf Aquitaine-Braspetro consortium was awarded a contract on block E in the northern part of Guatemala. In early 1982 a Texaco-Hispanoil-Braspetro consortium obtained a contract on block L, south of blocks D and E.[3]

Example of Contract Conditions Based on an Actual Contract

Actual contracts negotiated through 1982 have followed the model contract of January 1978. However, the provisions with respect to minimum drilling and investment requirements and state participation differ somewhat with the contract area and according to competitive bidding. The conditions stated in the following paragraphs are taken from one of the contracts negotiated in 1978.[4]

1. During the first three years of the contract life, the contractor agrees to drill at least two exploratory wells up to a maximum depth of 4,000 meters or until the geologic basement is found, whichever comes first. The first well must be drilled within 30 months of the date the contract goes into effect. The contractor also agrees to drill an exploratory well in each of the fourth, fifth, and sixth years of the contract life up to a depth of 3,500 meters, or until the geologic basement is found.

2. The contractor is required to guarantee the minimum investment for the first three years as follows: Q0.5 million for the first year; Q1.0 million for the second year; and Q10.0 million for the third contract year. The entirety of this

"Guatemala Block Goes to Texaco Group." *Oil and Gas Journal* (August 9, 1982) p. 71.

[4]The company holding the contract has requested that its name be kept confidential.

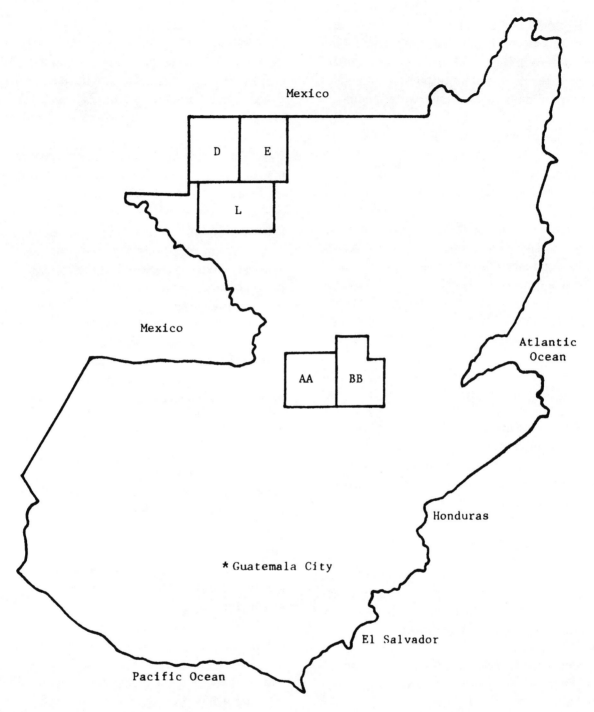

Contract holders:

        AA - Hispanoil Consortium
        BB - Getty Consortium
        D - Texaco-Amoco
        E - Hispanoil Consortium
        L - Texaco-Hispanoil-Braspetro

Figure 7-1. Guatemala Service Contracts

minimum investment must be guaranteed by the contractor before the contract is signed. The contractor is obligated to guarantee a minimum investment of Q4.5 million during each of the fourth, fifth, and sixth years of the contract life. These investments must be guaranteed before commencement of each of the yearly periods, but if the contractor relinquishes the entirety of the exploration area before the commencement of any yearly period, and the relinquishment has been approved by the government, the contractor shall be relieved of his responsibility to guarantee minimum investments corresponding to subsequent years.

3. The contract area measures 197,036 hectares. The contractor must relinquish the following percentages in the original contract area: (a) 20 percent before termination of the third year of the contract; and (b) another 30 percent before termination of the fifth year. The contractor is entitled to choose and retain for each commercial-size discovery one or more blocks not larger than 10,000 hectares within the contract area for exploitation. In special cases a larger block may be selected. The sum total of all blocks chosen and retained may not exceed 50 percent of the original contract area.

4. When the contractor has chosen one or more blocks for inclusion in the exploitation area, the exploration stage begins for each block on the date of choosing. The contractor is required to present and carry out within the contract area an exploitation program approved by the government for obtaining the most efficient recovery of the hydrocarbons. The contractor may relinquish one or more blocks of the original area without terminating the contract.

5. The state's share in the production of crude oil is (a) 58.15 percent of gross crude production from 1 to 50,000 bpd; (b) after applying the percentage specified for the first 50,000 bpd, the state's share is 61.16 percent for amounts exceeding 50,000 bpd, but not more than 100,000 bpd; and (c) after applying the percentages specified above, the state's share is 65.0 percent for amounts exceeding 100,000 bpd. This production split is somewhat more favorable to the contractor than that provided in the 1978 model contract.

6. If marketable natural gas is produced, the state's share is (a) 55 percent of gross production from 1 through 8 million cubic meters (cu.m.) per day; (b) 65 percent for amounts between 8 million to 16 million cu.m. production per day; and (c) 75 percent for amounts exceeding 16 million cu.m. per day.

7. The contractor is committed to carrying out a training and scholarship program for qualifying Guatemalan personnel amounting to Q125,000 yearly while no commercial-size discovery has been made; and of Q350,000 yearly after a commercial-size discovery has been determined and confirmed.

8. The contractor is committed to indemnify the landowners or any person to whom damages are caused as a result of petroleum exploration or exploitation activities.

9. The contractor is committed to build highways or improve existing ones during the exploration stage under specifications agreed with the government at a cost determined according to the following scale on the basis of the number of hectares constituting the exploitation area on the first day of each contract year: first year, none; second year, Q1 per hectare; third year, Q3 per hectare; fourth year, Q6 per hectare; fifth year, Q10 per hectare; and sixth year, Q20 per hectare.

10. Within the first three years of the exploration stage, the contractor is required to build a school and a hospital at a cost of at least Q300,000 and to pay for educational staff services, for services of at least one permanent doctor, and for services of the required paramedical personnel.

11. The government shall supervise through a committee the petroleum operations covered by the contract.

12. The contract shall lapse automatically if the contractor does not find exploitable hydrocarbons in commercial quantitites at the end of the sixth year of the contract life. The contractor has the right to terminate the contract at any time on condition that he pay the state the amount of the outstanding obligations which are applicable at the time of termination.

13. At the expiration of the contract, all buildings, fixed installations, machinery and equipment, and any other property that may be a part of the petroleum operations are to be transferred to state ownership at no cost, in good working condition, and free of any encumbrances or limitations.

14. The state reserves to itself all matters relating to the construction and operation of refineries, pipelines, and transportation media for hydrocarbons that may be discovered in the contract area. The contractor is entitled to utilize on a nondiscriminating basis any government transportation facility. However, the contractor may submit to the government a program for the development of new transportation facilities which the contractor may operate. The state reserves the right to use such facilities or to acquire them.

Criticisms of the Model Petroleum Contract

The following criticisms have been expressed regarding the 1978 model contract by representatives of companies that have negotiated contracts, by representatives of companies that failed to bid, by a leading Guatemalan petroleum consultant, and by a prominent Guatemalan tax accountant.

1. The Guatemalan income tax on petroleum production may not be creditable against the U.S. corporate income tax. Although three U.S. companies have accepted contracts, a number of other U.S. petroleum companies that were interested in Guatemalan production—including Exxon, Marathon, Occidental, Phillips, Shell, Texas Pacific, Union Oil—did not bid, reportedly in large measure because of the tax arrangements.

2. The exploration areas, 200,000 hectares or less, are regarded as too small, especially if a company is required to drill as many as five wells on half the area, as is the case with the Getty contract area.

3. Contract terms do not provide sufficient rewards in view of the risks, and there is virtually no room for negotiation.

4. The minimum well drilling and investment conditions are too high in advance of initial exploration, and the bid acceptance tax of Q1 million, together with the fees for making bids, are regarded as exorbitant.

5. The steeply graduated state participation rate discourages exploration and drilling for increased production in marginal areas.

6. The 2 percent petroleum export tax and the 2 percent stamp tax are both burdensome and uncertain in application. (The companies do not know whether they will be exporting and there are unresolved questions relating to the transactions to which this fiscal stamp tax will be applied.)

7. Some company officials regard the minimum infrastructure, education, hospital facilities, and training expenditure requirements to be excessive.

8. There are no plans for the utilization of gas should wells be found to produce gas rather than petroleum.

These are serious criticisms and help to explain why so few international companies took part in the bidding for the initial contracts. Indeed, in the case of U.S. companies that are uniquely affected by the income tax arrrangements, one is led to wonder why they bid at all.

There are different views regarding the outlook for finding oil in Guatemala, and some geologists are far more optimistic than others. Evidently the companies that bid on the contracts were quite optimistic, and one representative of an American company stated that his company decided to bid in spite of the obstacles in order to obtain a foothold in the petroleum operations of the country with the expectation that the tax law would be changed in the future in order to attract more companies. Another U.S. company official stated that unless it becomes possible to credit Guatemalan taxes against U.S. tax liabilities, his company will either have to shut down operations under the contract or sell its investment to a non-U.S. company. As of the time of writing, no change had been made in the tax provision, but nevertheless Texaco was a party to a new contract in 1982.

Activities of Petromaya and Centram

Petromaya drilled five wells and produced 5,000 bpd in 1981. It has built a 200-km pipeline from its producing field in the north to Port Santo Tomas on the Caribbean at an estimated cost of $30 million. The pipeline capacity is 15,000 bpd (with an ultimate capacity of 60,000 bpd) and will also go through the Getty contract area. Petromaya officials are fairly optimistic regarding the potential for their concession areas and believe that their areas are more promising than those along the Mexican border.

Shenandoah Oil withdrew from Petromaya and until 1982 it was owned 70 percent by Basic Resources International (incorporated in Luxembourg in 1981) and 30 percent by Elf Aquitaine. In 1982 Basic Resources was reported to have sold its interest in the joint venture.[5] In August 1980, Petromaya's original concession was terminated and replaced by a contract that conforms to the 1978 Petroleum Code. The new contract contains the following key points:[6]

1. The size of the operations area is reduced from 400,000 to 200,000 hectares.

2. The term will be no more than twenty-five years, as opposed to the forty-year term of the concession.

3. Ten wells will have to be drilled in six years and Q50 million will have to be spent during that time.

4. The government's participation is set according to the following schedule:

from 0 to 15,000 bpd	55%
from 15,001 to 30,000 bpd	60%
from 30,001 to 50,000 bpd	65%
from 50,001 to 100,000 bpd	70%
from 100,001 on	75%

Centram, owned originally by Inco (80 percent) and Hanna Mining (20 percent), held a concession dating from the mid-1960s. The concession was near Lake Izabel, where evidences of petroleum were found while the two companies were exploring for their nickel mine, Exmibal. Hanna withdrew from Centram and

the concession was then owned 87.5 percent by Inco and 12.5 percent by Canadian Superior. Some $26 million was spent on the concession over a fourteen-year period, including two offshore wells drilled in the Gulf of Honduras, but nothing of commercial value was found. Centram also farmed out one area of the concession to Arco and another to Allied Chemical, both of which conducted seismic work, but neither of the firms decided to join with Centram in a joint venture. The concession was forfeited in April 1980 for failure to carry out the drilling program.

Recent Oil Discoveries

In early 1981, Hispanoil reported a discovery in block AA with reserves estimated to be at least 50 million barrels and one well has tested at 4,850 bpd.[7] In 1982 Texaco announced a discovery in block D, which it operates on behalf of itself and Amoco. The Texaco-Amoco area is 150 miles east of Mexico's Reforma area.[8]

Guatemalan Government Policy and Administration of Petroleum

The Guatemalan government has no operating responsibility in petroleum production and apparently does not expect to have any in the future. The official responsible for oil and gas is the secretary of mines, hydrocarbons and nuclear energy, with ministerial rank but not under any of the regular ministries. The secretary reports directly to the president and is assisted by the National Petroleum Commission and by the assistant secretary of mines, hydrocarbons, and nuclear energy, who has considerable responsibility for petroleum policy. Both technical and policy advice are provided by the United Nations Development Programme (UNDP) through a Canadian advisory group, and some of the private company representatives believe that the

[5]*Wall Street Journal* (August 17, 1982) p. 13.
[6]Information obtained from the U.S. Embassy in Guatemala.

[7]*Petroleum Economist* (May 1981) p. 216.
[8]*Petroleum Economist* (August 9, 1982) p. 71.

government's rigorous contract terms reflect UNDP advice.

Conclusion

For a high-risk country with little petroleum discovered, Guatemala's contract terms are poorly designed to attract exploration or to maximize the government's proportion of the economic rents. The high front-load outlays, including the Q1 million bonus payment and minimum exploration outlays over a three-year period re- gardless of the results of the seismic surveys or the initial exploration drilling, have minimized the attractiveness to petroleum companies. Only a few large companies have made bids and there has been little competitive bidding. On the other hand, if a large oil reservoir were found, the state might be recovering a smaller share of the economic rent than it would be possible to ob- tain under a different set of contract terms. Moreover, the graduated share of the output going to the government based on total output from each contract area is likely to work against the efficient development of the petroleum reserves.

8

The Production-Sharing Contracts of Malaysia, the Philippines, and Egypt

Malaysia, the Philippines, and Egypt employ PSCs, but there are substantial differences in the contract provisions and in their fiscal implications. Egypt and Malaysia each have reserves of about 3.3 billion barrels, but Egypt is producing at twice the rate of Malaysia, and Egypt's output and reserves have been expanding rapidly in comparison with those of Malaysia. Egypt has been much more successful in attracting foreign companies for exploration than Malaysia.

Malaysia

Early History

The Malaysian Federation had no unified petroleum legislation during the initial concession period until the Petroleum Mining Act of 1966 which, together with the 1968 Petroleum Rules, established a system of exploration licenses calling for compulsory surrender of 50 percent of the original concession area after five years and 75 percent after ten years. The 1967 Petroleum Income Tax Act provided for a 50-50 profit-shar-

ing arrangement based on the prevailing OPEC model. In the 1968–69 period, the Malaysian government granted *provisional* exploration concessions to Exxon, Continental Oil (Conoco), Southeast Asia Gulf, Mobil Oil Malaysian, and Amoco Malaysian Petroleum. These grants were contingent upon future settlement of final terms, which included a minimum of 15 percent government participation.

Developments Since 1974

The Petroleum Development Act of 1974 established a national petroleum company, Petronas, and required the conversion of existing concessions to either production-sharing or joint-venture agreements with Petronas. This led to a dispute with the foreign companies, particularly with Esso, but in 1976 PSCs were negotiated with Shell and Esso.[1] Conoco withdrew in 1978. Some of the

For a good historical summary, see Office of International Affairs, "Malaysia," in *Role of Foreign Governments in the Energy Industries* (Washington, D.C., U.S. Department of Energy, 1977) pp. 369–374; see also Corazon Morales Siddayao, *The Supply of Petroleum Reserves in South-East Asia* (Kuala Lumpur, Oxford University Press, 1980) pp. 91–92; and "Malaysia," *Petroleum News* (January 1981) p. 36.

features of the 1976 contracts include the following:

1. A signature bonus of $1 million; a discovery bonus of $1 million for each commercial discovery; and a production bonus of $2 million when production reaches 4,650 barrels per quarter.

2. A 10 percent royalty based on official prices.

3. Recovery of costs up to 20 percent of gross revenue for oil and 25 percent for gas, with unrecovered costs carried forward.

4. The amount remaining after royalty and cost recovery is split 70-30 in favor of the government. (The split is 65-35 for gas.)

5. The contractor must pay Petronas 0.5 percent of the proceeds of the sale of contract cost oil and profit oil for use in a research fund.

6. The contractor pays a 45 percent tax on net earnings from his share of the crude, with bonus payments, royalties, export taxes, costs, and excess proceeds tax deductible for tax purposes.

7. The contractor must pay an additional tax equal to 70 percent of his proceeds above a base price, which is increased at an annual rate of 5 percent to reflect inflation. The 1978 base price was $12.70 per barrel.

8. Effective April 1980, oil exports became subject to a 25 percent crude oil export duty applied against the value of the contractor's production share.

9. Duration of the contracts is twenty years, with possible four-year extensions for oil and fourteen years for gas.

Evaluation of Malaysian PSCs

An analysis of Malaysia's contract terms as summarized above suggests that they are among the least attractive of those employed by non-OPEC countries. The signature and production bonuses; the 10 percent royalty; the relatively low 20 percent recovery costs; the 70-30 split in the government's favor; the 45 percent tax on earnings plus the additional tax of 70 percent of the con-

tractor's proceeds above the base price; and the 25 percent export duty imposed in April 1980 combine to make the government's take virtually confiscatory, except for relatively large, low-cost fields. The results of the simulations of operations under Malaysian contract terms shown in table 4-4 indicate virtually zero IRR for high-cost/medium-volume and low-volume hypothetical fields, and the IRR for a medium-cost/medium-volume field is only marginally acceptable without taking into account the risk-corrected IRR.

Simulations of the results of operations under Malaysian contracts (as of mid-1980) under alternative development cost and price assumptions by W. J. Levy Consultants Corp. yield comparable results. In the case of a relatively small field containing 15 million barrels and development costs per daily barrel of average production of $21,900, the Malaysian terms provide no incentive for field development, since the government's take is virtually 100 percent; even for large fields containing 250 million barrels and development costs per daily barrel of average production of $12,351, the IRR is barely marginal and uneconomic at higher cost levels. For a median cost/price case for the 250 million barrel field, the government's take on the contractor's equity share is 95.6 percent compared with 87.1 percent for the Indonesian fiscal system, 82.8 percent for the Norwegian system, and 80.2 percent for the U.K. system.[2]

It is not surprising that Malaysia has attracted relatively little interest on the part of international petroleum companies. The first contract to be signed since the forced renegotiation of the Esso and Shell contracts in 1976 was a joint-venture, production-sharing contract between a consortium of British Petroleum and Oceania, on the one hand, and Petronas, on the other, in May 1980. No additional contracts were signed until 1982 when a joint-venture contract was concluded between Petronas and the French government-owned enterprise, Elf Aquitaine.[3] This contract provided for cost recovery of 30

[2]See W. J. Levy Consultants Corp., "Comparative Analysis of Exploration Arrangements in Selected Countries," (New York, November 1980) tables I-1 and II-3.

[3]A private Malaysian firm, Delcom Services, has a 4 percent equity share.

percent of gross oil production and 35 percent for gas. The remaining oil after payment of the 10 percent royalty is split 70-30 between Petronas and the joint venture. Under this contract, 15 percent of the equity is held by Petronas. Elf Aquitaine paid a signature bonus of $7 million and agreed to spend $50 million during the first five years of exploration. Elf Aquitaine will bear all exploration costs until the first four wells are drilled.[4] The larger recovery allowance could improve the expected IRR on large fields, but might not render the development of small fields economically feasible.

The Philippines

Early History

The Philippine Petroleum Act of 1949 proclaimed state ownership of all deposits of petroleum or natural gas, with private exploration and exploitation permitted on the basis of more or less standard concession contracts and remuneration to the government based mainly on a royalty of 12.5 percent. A number of concessions were granted and by 1964 over 250 exploratory wells had been drilled with no commercial discoveries. There followed an inactive period until the passage of the Oil Exploration and Development Act of 1972 (PD 87). This act established the "service contract" system in place of the concession contract for private exploration and development of petroleum. The Philippine service contract is similar to the Indonesian production-sharing contract, but important differences exist. For example, the contracts with the private petroleum companies are negotiated with the Bureau of Energy Development (BED) which operates under the Ministry of Energy, rather than with the state petroleum enterprise, the Philippine National Oil Company (PNOC), which was chartered in 1973.[5]

The Philippine Service Contract

The Philippine service contract, which is essentially a PSC, has evolved during the period since 1972. The basic features and certain important changes are described in the following paragraphs.

The contract provides that all necessary services, technology, and financing in both the exploration and production periods be supplied by the contractor. The maximum contract area is 750,000 hectares onshore and 1 million hectares offshore. The initial exploration period is for seven years, extendable to ten years. If petroleum is discovered, production is permitted for twenty-five years beyond the exploration period, plus additional extensions not exceeding fifteen years. Twenty-five percent of the area must be relinquished after the first five years and, in the case of an extension, an additional 25 percent relinquished at the end of the seventh year. In the production stage, a contractor may retain the producing area plus 12.5 percent of the original area for exploration. Rentals must be paid on the acreage held.

There are minimum expenditure requirements per hectare per year which vary during the exploration period and between onshore and offshore exploration. These amounts are negotiable, but a recent agreement establishes the minimum expenditures onshore at $3 per hectare during the first five years and $9 thereafter; for offshore, $3 per hectare during the first and second years, $6 from the third to fifth years, and $18 from the sixth to tenth years.[6] The contractor also may be required to drill a specific number of wells each year, but in 1980 contractors were offered a "seismic option," relieving them of specific drilling requirements if seismic surveys did not show that exploratory drilling was warranted.

More recent contracts provide for the payment of signature, discovery, and production bonuses. The latter are on a negotiable sliding scale beginning with the declaration of commercial discovery and rising with the output in terms of barrels per day. The cost recovery

[4] "Petronas Improves Cost Recovery Provisions," *Oil and Gas Journal* (November 29, 1982) p. 48; see also, *Petroleum News* (January 1983) p. 28.

[5] For a summary of PD 87 see Siddayao, *Supply of Petroleum Reserves,* pp. 93-94; see also, *Petroleum News* (January 1981) p. 54 for more recent developments.

[6] Siddayao, *Supply of Petroleum Reserves*, p. 95.

allowance in the contracts has varied from 55 to 70 percent, with allowance for forward transfer of additional operating costs, but more recently a 60 percent cost recovery allowance appears to be the standard. An additional allowance of 7.5 percent of gross proceeds is provided for contractors with Philippine partners. Assuming a contractor has a Philippine partner, the total allowance before the production split is 67.5 percent of gross proceeds, leaving net proceeds of 32.5 percent to be divided. Most contracts provide for a 60-40 split of net proceeds in favor of the government, but some contracts provide for a 70-30 split for production in excess of 75,000 bpd. Until recently, the Philippine government's share of net proceeds automatically included income taxes payable to the government by the contractor. However, as has already been noted, these taxes can no longer be credited against U.S. tax obligations. Therefore, in 1981 an amendment to the existing service contracts was made providing for a "production allowance" which increases the contractor's share of the net proceeds, but this allowance is actually paid back to the government in the form of income taxes and thus is considered as a tax eligible for credit by the IRS.

Assuming a 60 percent cost recovery allowance and a 60-40 profit-sharing ratio, the proportion of gross proceeds represented by the profits share of the contractor and the government may be shown as follows (in percent):

Gross proceeds	
Philippine partnership incentive allowance	7.5
Cost recovery allowance	60.0
Net proceeds	32.5
Production allowance	16.0[a]
Adjusted net proceeds	16.5
Sharing (60-40)	
Service contractor	6.6
Government	9.9

[a]Actually paid back to the government in the form of direct taxes.

On the basis of the above analysis, the service contractor receives about 20 percent of the net proceeds while the government receives 80 percent. However, this does not take into account bonus payments, which could reduce the contractor's share significantly, especially for relatively small discoveries. Unfortunately, the bonus payments are confidential and I have been unable to obtain information on their general magnitude.

There is no royalty, and petroleum sold in the domestic market is valued at a price determined by the BED, which is based on the OPEC price. Although the contractor has the right to export his portion of the petroleum produced, this is limited by the requirement to meet Philippine domestic demand according to a ratio of the contractor's output to total production in the Philippines. Since it will be many years (if ever) before enough petroleum will be produced to satisfy the domestic demand, it seems unlikely that private companies will be able to export crude for some time.

Imports of machinery, equipment, spare parts, and materials required for petroleum operations are exempt from tariffs and other taxes, provided they are not available from domestic sources. Filipinos are to be given preference in employment, and the contractor is required to provide schooling and training to Filipino personnel after commercial production begins.

Disputes between the BED and the contractor relating to the contract, which cannot be settled amicably, are to be settled by arbitration. If the parties cannot agree upon a third arbitrator, one shall be appointed by the president of the International Chamber of Commerce.

A service contract with a consortium headed by Cities Service signed in November 1980 provided for a commitment to drill five wells over seven years and a profit split of 67.5 to the government and 32.5 percent to the contractors. A service contract signed in August 1980 by a Philips Petroleum–Shell joint venture calls for a minimum expenditure of $31 million and the drilling of five wells over a seven-year period. No information on the signature or production bonuses for either of these contracts is available.[7]

[7]Petroleum News (January 1981) p. 52.

Contracts and Commercial Discoveries

By the end of 1980, 34 service contracts had been signed with the BED, including several by the government enterprise, Pnoc-Exploration Corporation (Pnoc-EC), incorporated in 1976. However, some 15 contracts had been relinquished. In 1981 the principal companies or consortia leaders conducting operations were Amoco, Cities Service, and Philips Petroleum. Most of the service contracts are joint ventures involving several petroleum companies. During the period 1972–77, 65 exploratory wells had been drilled, but not until 1977 were there any commercial discoveries, when two of the 12 wells drilled that year were successful. There were several additional discoveries during 1978–80 and Cities Service made an important strike in 1981. All the discoveries, save one onshore made by Pnoc-EC, have been in the Palawan Basin (offshore). Four wells began producing in the Nido oil field (NW Palawan) in 1979 (operated by Cities Service) with a peak output of 40,000 bpd in September 1979. In February 1980 the original well began pumping water and other wells in operation were temporarily shut down. These wells resumed production in April 1980, but their output gradually declined from 14,000 bpd to only 3,000 bpd at the end of 1980.[8] Two other fields, Cadlao and Matinloc, in the same offshore region were discovered by Amoco and Cities Service, respectively. The Cadlao field began commercial production in September 1981 with an output target of 9,000 bpd, while the Matinloc field came on stream in July 1982 at 13,500 bpd, bringing Philippine production to 25,000 bpd at the end of 1982, about 1.2 percent of the country's consumption. BED estimates Philippine crude output at 26,000 bpd by the mid-1980s.[9] Exploration activities by Pnoc-EC have been mainly confined to onshore areas, while the private companies have been operating almost entirely offshore.

Despite the disappointing results of petroleum exploration and production in the Philippines, petroleum companies continue to be attracted to the area, and a substantial amount of exploration is taking place. In addition to the Cities Service and Amoco consortia, joint ventures headed by Phillips Petroleum, Total, British Petroleum, Pecten (Royal Dutch Shell), Arco, and St. Joe Petroleum, among others, continue to be active. Some companies have moved in and out of the area several times over the past five years, despite the evidence from actual discoveries thus far that future finds are likely to be moderate at best.

Evaluation of Philippine Contracts

The Philippine offshore areas are characterized as high risk in terms of discovery probability, high drilling costs, and low volume of recoverable oil in the fields. The reservoirs discovered in the Philippines, at least through 1981, have averaged only 10 million barrels each. Under such conditions, it is especially important for Philippine contract terms to provide a relatively high expected IRR for low-volume/high-cost fields. Except for the bonus provisions, which mainly take the form of progressive production bonuses, the provisions of the Philippine contracts such as the relatively high cost recovery allowance, a heavy reliance on the corporate income tax for the government's take, and the absence of a royalty together suggest that the Philippine system is more likely to be progressive than regressive.[10] However, the combination of 60-40 (or higher) production split and the 60 percent Philippine income tax plus an additional 7.5 percent tax on profits remitted abroad by foreign contractors may make the expected IRR on relatively small fields uneconomic. Press reports indicate that companies have stated that more liberal contract terms are necessary if an adequate level of exploration is to be undertaken. In 1981 the World Bank conducted a study of the Philippine energy sector and made some recommendations to the

---

[8]See "Philippines: Further Setbacks to Oil Production." *Petroleum Economist* (February 1981) p. 76.

[9]"Philippines: Second Oilfield Onstream," *Petroleum Economist* (February 1981) p. 405; and "Philippines: Exploration Hopes Offshore Palawan," *Petroleum Economist* (December 1982) pp. 507–508.

[10]I do not have the cost data necessary to undertake a simulation analysis of the effects of the Philippine fiscal system on net returns to the contractors for fields of differing quality.

government for liberalizing contracts in order to increase incentives for drilling.[11] Legislation liberalizing cost recovery was introduced in late 1982.[12]

Egypt

The Egyptian system formally provides for a 50-50 joint venture with an oil company and the state enterprise, Egyptian General Petroleum Corporation (Egpc), following a commercial discovery by the company under an exploration concession. However, the arrangement is essentially a PSC. The exploration concession provides for a specified expenditure commitment, a signature bonus which may range from $1 to $5 million, and production bonuses graduated with the output of the field. During the production period, 30 percent of annual production is allowed for cost recovery, but capital expenditures can only be recovered on a straight-line basis over an eight-year period. Any excess of the cost oil allowance over actual eligible cost must be returned to Egpc. The company's share of the profit oil ranges from 15 to 20 percent, depending upon the production rate, with an 80-20 split in favor of Egpc on the first 90,000 bpd, rising to 85-15 on production exceeding 140,000 bpd. Egpc pays all taxes and royalties, but contributes nothing toward production costs. (The implications for U.S. tax liabilities are discussed in chapter 11.)

As was noted in table 4-4, the Egyptian system is progressive in terms of the government's take of net profits, declining from 87.5 percent for the low-cost/high-volume fields to 81.2 percent for the high-cost/low-volume fields. However, the IRR in real terms is only satisfactory for the low-cost/high-volume and medium-cost/

medium-volume fields. The progressivity of the government's take is explained entirely by the fact that the cost recovery oil which exceeds that actually required to recover costs must be returned to Egpc, and the excess of the cost recovery allowance over that actually required tends to decline with the quality of the field. Unlike most non-OPEC LDCs, Egypt has several giant fields and Egyptian field discoveries tend to be considerably larger than those of most non-OPEC LDC oil-producing countries. This may explain why Egypt has been able to attract a larger number of petroleum companies to explore for oil despite the fact that expected IRR on relatively small, high-cost fields is unattractive.[13]

Conclusions

Of the three countries whose PSC systems are described in this chapter, the resource potential in terms of known recoverable oil and giant fields is highest for Egypt, followed by Malaysia and the Philippines.[14] The contract terms of Egypt and Malaysia are substantially more severe than those of the Philippines, particularly with respect to their impact on returns on smaller fields, but Egypt and the Philippines have been more successful than Malaysia in attracting foreign petroleum companies to exploration. Malaysia's contract terms are uneconomic for the high-cost/medium- and low-volume fields, and are only marginally attractive for the medium-cost/medium-volume fields. In fact, according to the Walter J. Levy group's simulations, Malaysia's fiscal system is attractive only for a low-cost/high-volume field. Although I lack the data for a simulation of the Philippine contract terms, the terms appear to be better suited for a country with high-cost/low-volume fields because of the progressivity of their effects on the government's take.

[11]The World Bank report is confidential, but some of its recommendations are summarized in "Philippines," *Petroleum News* (January 1982) pp. 40–41.

[12]*Petroleum News* (January 1983) p. 47.

[13]See Kemp and Rose, "Four Oil Systems."

[14]Nehring, *Giant Oil Fields*, pp. 32–33.

9

The Service Contracts

As stated in chapter 3, the term "service contract" covers a variety of arrangements, from the pure service contract under which the contractor performs specific tasks for a specified sum of money, to the risk-service contract that has features of the modern concession agreement and the production-sharing contract. There are also some service contracts that involve only a limited amount of risk on the part of the contractor. The Argentine government has negotiated a whole range of types of service contracts, from the pure service contract to the risk contract; the Brazilian government negotiates only risk-service contracts. To call an arrangement in which the contractor puts up all the capital at risk and is rewarded solely on the basis of the outcome of a risky investment a service contract is a clear misuse of the term, but this usage has evolved in large part because governments have for reasons of public policy wanted to avoid any association with the traditional concession agreement in dealings with private petroleum companies.

The Argentine Petroleum Contracts

Brief History of Argentina's Petroleum Production Policies

Since oil was first discovered in Argentina in 1907, Argentina's petroleum policy has shifted every few years with changes in government leadership.[1] The government's petroleum enterprise, Yacimientos Petroliferos Fiscales (YPF), was organized in 1922 and given the right to enter all phases of the oil business, from exploration to transportation and marketing. To implement these rights, YPF received from the government certain oil lands known as state reserves as well as capital funds, so that from 1922 to 1935 YPF operated in competition with private companies in Argentina. Before 1922, several private companies received concessions to undertake exploration and development. Among

---

[1] For a history of Argentine petroleum policy to 1968, see Edwards, "The Frondizi Contracts," in Mikesell, *Foreign Investment*, chapter 7.

these were Shell, Standard Oil of New Jersey, and the Argentine company, Astra. By 1934 private companies were contributing about 60 percent of total Argentine production. However, in 1935 the state reserves were increased and concessions of private companies were limited, with all new exploration and development reserved for YPF. Although private companies continued to operate, they were producing about half as much oil in 1957 as they were in 1935. Meanwhile, YPF had only managed to increase its production from about 26 percent of total consumption in 1935 to 33 percent in 1957, and between 1947 and 1955 oil imports doubled.

The Argentine constitution of 1949 specified that all underground resources, including petroleum, were the inalienable property of the national government. The government took the position that it was unconstitutional to grant further concessions to private companies, but without new concessions, the private companies already operating were unlikely to increase production. However, a law passed during the presidency of Juan Peron in August 1953 (Law 14222) established conditions for admitting foreign capital to the oil industry, and just prior to the overthrow of the Peron government a concession contract with mixed Argentine–U.S. company participation was approved by executive decree and submitted to the Argentine legislature. The contract was subject to vigorous debate and was still not approved by the legislature when the military coup of September 1955 removed Peron from office.

Nothing much happened thereafter until Arturo Frondizi took office as president in May 1958. During his election campaign Frondizi promised Argentine self-sufficiency in oil as soon as possible, but at the time he advocated the job should be done entirely by YPF. He quickly abandoned this position and between 1958 and 1961 the Argentine government negotiated a large number of contracts with private companies—both foreign and domestic (table 9-1).[2] The contracts were of three types: drilling, develop-

ment, and exploration/development. Drilling contracts provided that each company should drill a specified number of wells in a specified area. Payment terms were stipulated on the basis of fixed amounts per meter drilled and per hour spent on completion of the wells. Once the wells were completed, the company's responsibility ended and actual production was undertaken by YPF. Under the development contracts, private companies undertook to develop for YPF areas in which oil reserves were known to exist. Few risks of exploration were involved and the companies received payment in accordance with the amount of crude oil delivered to YPF from developed wells. In the exploration/development contracts, the work was to be done for YPF in new and semiproved areas and the risks of exploration fell on the contractors. Payment was conditional on oil being found in commercial quantities. In the case of the contract with Royal Dutch Shell of December 1958, payment was to be made in crude oil while for most of the others payment was made in money.

In physical terms the work done under the contracts was very successful. Total production of oil in Argentina rose from 97,800 bpd in 1958 to 266,000 bpd in 1962, while crude oil imports fell from 130,200 bpd to 20,600 bpd. Nevertheless, the contracts were very unpopular politically and were an important factor in Frondizi's being removed from office by a military coup in March 1962.

The elections of 1963 gave the presidency to Arturo Illía, who in November 1963 made good on his campaign promise to cancel all three types of contracts. With the return of full control over the petroleum industry to YPF, production declined while demand continued to rise, and the number of wells drilled fell from 1,639 in 1961 to 466 in 1964. In 1965 YPF negotiated some drilling contracts and the number of wells drilled increased to 660 in 1966.

After General Juan Carlos Onganía succeeded Illía as president of Argentina in June 1966, Argentina's petroleum policy changed again in favor of private investment, and a new hydrocarbons law, Law 17319, was enacted in June 1967. The most striking feature of the new legislation was that it permitted the government to grant oil

[2]The foreign companies included ENI, Kerr-McGee, Standard Oil of Indiana, Cities Service, Esso, Royal Dutch Shell, Marathon and Union Oil. See accompanying table 9-1.

Table 9-1. Argentina: The Frondizi Oil Contracts 1958–62

Company	Date of contract	Period	Area	Payment terms
Drilling				
Societá Azionaria Italiana Perforazioni Montaggi del Grupo E.N.I. (SAIPEM) [a subsidiary of Ente Nazionali Idrocarburi (E.N.I.)]	April 1959/August 1960	600 wells by 1963	South Flank (C. Rivadavia)	$20.50 per meter drilled and $50 per hour spent completing (62% in lire, 38% in pesos)
Southeastern Drilling Company of Argentina, S.A.	April/August 1959	1,000 wells in 4 years	South Flank (C. Rivadavia)	$19 per meter drilled and $50 per hour spent completing (60% in dollars, 40% in pesos)
Kerr-McGee Oil Industries, Inc.	April 1959 (revised, see below)	500 wells in 3 (later amended to 4) years	South Flank (C. Rivadavia)	$19.80 per meter drilled and $50 per hour spent completing (60% in dollars, 40% in pesos—later amended to 75% and 25%, respectively)
The Transworld Drilling Company (Kerr-McGee)	August 1963	100 wells (of Kerr-McGee's original 500)	South Flank (C. Rivadavia)	$15.84 per meter drilled and $50 per hour spent completing (75% in dollars, 25% in pesos)
		350 wells (all 850 wells under 1959/63 contracts to be completed within 6½ years from August 1959)	South Flank (C. Rivadavia)	$11.88 per meter drilled and $47.50 per hour spent completing (75% in dollars, 25% in pesos)
Development				
Pan American International Oil Company (Standard of Indiana)	July 1958	15 to 20 years	4,000 km² in C. Rivadavia	$10 per m³ oil delivered to YPF (60% in dollars, 40% in pesos)
Argentine Cities Service Development Co. [formerly Carl M. Loeb Rhoades & Co.] (Cities Service, 63⅓%; Signal Oil & Gas, 20%; Union Oil & Gas of California, 16⅔%)	July 1958	20 years	4,860 (later reduced to 1,860) km² in C. Rivadavia and 480 km² in Mendoza	65 to 70% of foreign exchange saving based on prices of local and imported crude
Tennessee Argentina, S.A. (Tennessee Gas Transmission)	April 1959	25 years	14,000 km² in Tierra del Fuego	$11.15 per m³ oil and $2 per 1,000 m³ gas delivered to YPF
Astra Compañía Argentina de Petróleo, S.A.	December 1961	20 years	60 km² in Santa Cruz	$8.967 per m³ oil delivered to YPF (in pesos)
Compañía Argentina para El Desarrollo de La Industria del Petróleo y Minerales, S.A. (CADIPSA)	February 1962	20 years	64 km² in South Flank (C. Rivadavia)	$10 per m³ oil delivered to YPF up to 2 million m³; $9.5 per m³ third million; $9.0 per m³ fourth million; $8 per m³ in excess of 4 million m³

Table 9-1. Argentina: The Frondizi Oil Contracts 1958–62—continued

Company	Date of contract	Period	Area	Payment terms
Exploration/Development				
Esso Argentina, Inc. (a subsidiary of the Standard Oil Company of New Jersey)	December 1958	30 years	4,800 km² in Neuquen	$11–11.5 per m³ oil produced (in pesos)
Esso (revised)	September 1961	30 years (from December 1958)	1,892 km² of original area in Neuquen, and 14,194 km² in Neuquen, Mendoza, Rio Negro	
Shell Production Company of Argentina, Ltd. (Royal Dutch/Shell)	December 1958	30 years	30,000 km² in Rio Negro and Buenos Aires	In crude oil. Shell receives crude equivalent to its capital amortization and operating costs, YPF gets 10% of balance of crude output up to maximum value of £1½ million. Crude then shared equally by Shell & YPF
Shell Production (revised)	September 1961	30 years (from December 1958)	10,000 (later reduced to 1,000 km² of original areas in Rio Negro and Buenos Aires and 20,000 km² in Rio Negro, La Pampa, Neuquen, Mendoza)	
Marathon Petroleum Argentina, Ltd., and Continental Oil Company of Argentina (working jointly)	June 1961	40 years	67,140 km² in Santiago del Estero and Tucuman	$11.75 per m oil delivered to YPF at wellhead, or $14 if at San Lorenzo refinery (60% in dollars, 40% in pesos)
Union Oil Company of California	September 1958	22–27 years	16,000 km² in Santa Cruz and Chubut	$12 per m oil delivered to YPF (60% in dollars, 40% in pesos)

From Mikesell. *Foreign Investment in Petroleum and Mineral Industries*, pp. 158–159.

concessions. Private companies and mixed companies as well as YPF were permitted to engage in exploration, development, transportation, and marketing of oil and gas, although the deposits themselves continued to be the property of the state. This law is still on the books although no concessions have been granted in recent years. The concessions held by Cities Service and Amoco were converted into service contracts and these companies continue to work the contract areas today. For example, in 1978 Cities Service drilled the first of three wells of a 30-well drilling pro-

gram in the Mendoza area. Esso and Shell also retain the refineries which they built prior to the Frondizi administration.

The return of Peron to power resulted in another reversal of petroleum policy. During the 1970s the bulk of Argentina's crude output was produced by YPF and private contractors operating in areas with proved reserves, with only a small amount being produced under the old concessions held by local private companies. In 1978 about 70 percent of the wells were operated by YPF and the remaining 30 percent by private contractors. Amoco and Cities Service were the most important foreign contractors from the standpoint of volume produced.

The average production of Argentine wells is only about 70 bpd. A large number must be drilled each year as the older wells become depleted, and by 1978 some 26,000 wells had been drilled since production was initiated, of which 6,000 were in operation. Crude production rose from 396,000 bpd in 1976 to 470,000 bpd in 1979, while imports declined from 60,000 bpd in 1976 to 34,000 bpd in 1979. In 1980 Argentina was producing about 92 percent of its petroleum requirements, but over half the output was from secondary recovery. However, in 1982 Argentina's proved reserves of 2.65 billion barrels were sufficient for only fifteen years of production at the current rate of 483,000 bpd, and output is almost sure to decline in the near future unless there is a substantial increase in reserves.[3]

Exploitation Contracts

Under the exploitation contracts in areas of known reserves and for secondary recovery, private companies make investment expenditures with their own funds and receive a price for their output based on the Argentine price of crude, which in 1979 ranged from $8.50 to $10.00 per barrel. The price is fixed by the government rather than YPF, with which the contracts are negotiated. Contractors are paid on the basis of output in accordance with an annual program. The price is adjusted periodically by a formula (which may differ for each contract) based on the Argentine wholesale price index for nonagricultural goods, the wholesale price index for imported nonagricultural goods, and the Argentine industrial wage index. The more recent exploitation contracts provide a price to the contractors which in 1979 was about $9.50 per barrel, whereas the older agreements under which Cities Service and Amoco are operating provide a price of only about half that amount. However, until recently taxes were lower under the Cities Service and Amoco contracts since YPF had been paying the corporate tax.[4]

Tenders for bids on exploitation contracts are issued by YPF from time to time and interested companies bid on them in terms of the number of wells that they agree to drill and the amount they intend to spend, or on the basis of the contract price adjusted by a formula. In 1979 two zones in the province of Jujuy along the Bolivian border were opened for bidding, and the contract was won by a consortium headed by Bridas, an Argentine company. A consortium of Cities Service, Esso, Allied Chemical and four Argentine companies and an Amoco consortium also submitted bids. Bidding was apparently on a price basis and the Bridas bid was 16 percent more favorable to YPF than the Cities Service–Esso consortium bid. The local manager of one U.S. subsidiary suggested that this was due in part to the fact that Bridas as an Argentine company does not pay the 17.5 percent dividend withholding tax to which foreign companies are subject. Tenders were issued for several additional exploitation contracts in 1980.

1978 Risk Contract Law No. 21778

By 1978 it became clear that Argentina's petroleum output could not keep pace with the growing domestic demand unless there was a substantial increase in petroleum reserves. Since much of the Argentine mainland had been explored, the outlook for discovering large onshore fields was considered poor, and it was widely believed that Argentina's best chance of discovering additional reserves was in the offshore areas of the southern part of the country. YPF's limited offshore exploration efforts had not been successful, in large part because YPF lacked the technical capacity for such operations. Offshore drilling operations are also very expensive. It was, therefore, decided to invite private companies to negotiate risk contracts covering zones, either offshore or onshore, on which there had been little or no exploration.

The conditions for the negotiation of new risk contracts were set forth in the 1978 law, No. 21778, entitled "Risk Contracts for Exploration and Exploitation of Hydrocarbons." Contracts

---

[3]Argentina's current annual production is about 176 million barrels so that the production-to-reserves ratio is 1 to 15.

[4]These payments were prohibited by a 1979 Argentine law.

are awarded by competitive bidding in zones available for contracting that are announced from time to time by YPF. Exploration time may not exceed seven years for offshore operations and five years for land operations, in both cases counted from the date of legal validity of the contract, with the possibility of extending such terms for two additional years to evaluate a discovery. The time limits for development and production may not exceed twenty-five years from the date on which a commercial oil field is discovered. If natural gas is discovered for which there is no current market or transportation, a suspension of up to ten years may be granted.

Other important provisions of the risk-contract law are summarized as follows:

1. Foreign companies are required to have a domestic partner, but the domestic partner may own as little as 1 percent of the joint venture.

2. Contracting companies are subject to the general income tax regulations which provide for a graduated tax on net income, but there is no royalty on production.[5] There is provision in the risk-contract law for 100 percent amortization of the value of depreciable assets used during the exploration period, with such amortization to be used only against taxable profits arising from the contract. Tax losses in pesos may be escalated on the basis of the variation in the Argentine wholesale price index.

3. Imports required for execution of the contracts are exempt from payment of import or other duties, provided that the goods are not produced in the country efficiently and at reasonable prices and delivery dates.

4. Bidding on contracts may take the form of the agreed-upon number of wells to be drilled, or the amount of investment to be made during the exploration period, or on the basis of the K factor in the formula which determines the price to be received for the oil by the contractor from YPF. The first tender for offshore contracts announced in 1978 provided for bids in terms of the K factor, while the first tender in 1979 was on the basis of the number of wells and amount of expenditure. Where bidding is on the basis of the K factor, the number of wells and the minimum expenditure are established by YPF in the contract.

FORMULA FOR DETERMINING THE PRICE OF OIL RECEIVED BY THE CONTRACTOR. The price for petroleum to be paid by YPF to the contractor on delivery is a function of the world price of the relevant grade of petroleum, adjusted by the peso-dollar exchange rate; the Argentine wholesale price index for nonagricultural products; and a coefficient, K. According to the formula established by YPF, the lower the value of K, the lower the price that is paid. Depending upon the nature of the bidding, K may be established in advance by YPF or may be subject to bidding by the contractor. In either case K declines with the amount produced by the contractor in accordance with a fixed scale. For example, if K were fixed at 0.63 for production of up to 5,000 cubic meters[6] per day, for production between 5,000 and 10,000 cubic meters per day K would be 0.58; for production between 10,000 and 30,000 cubic meters per day K would be 0.50; for production between 30,000 and 60,000 cubic meters per day K would be 0.40; and for production in excess of 60,000 cubic meters per day, K would be 0.25. The price paid to the contractor is normally the K factor times the world oil price, but this is affected by other variables in the formula.

INITIAL BIDS UNDER THE RISK-CONTRACT LAW. The first tender under the Risk-Contract Law No. 21778 was for exploration and development of an offshore area covering 4,100 sq. miles just off the coast of Tierra del Fuego. Several groups presented bids, including Shell, Esso-Cadipsa, Gulf, Canadian Superior, and Chevron. The contract was won by a consortium which included Bridas and Arfranco (both Ar-

---

[5] In 1980 the Argentine corporate income tax was 33 percent. There was also a 17 percent foreign remittance tax, making a total tax of 50 percent on earnings of nonresidents. However, some foreign companies may not pay the foreign remittance tax.

[6] 1 cubic meter = 6.29 barrels of crude.

gentine), Deminex (West Germany), and Total Exploration (France). The contract requires that the Bridas group invest about $16 million, drill nine wells, and conduct 4,000 km² of seismic surveys during the first three years of the exploration period. Esso-Cadipsa actually proposed a larger volume of investment, but attached certain conditions and asked for clarification on certain points in the contract, which Esso believes resulted in its not obtaining the contract.

In September 1978 YPF tendered additional offshore blocks under the risk-contract law in the areas of Rio Gallegos and the Magellan Straits. The bidders included a group headed by Royal Dutch Shell (including Capsa and Petrolar), Bridas-Total, and Esso-Cadipsa. The Shell group, which won the bid for the two offshore zones, bid as follows: $42 million in expenditures during the first period; $16 million for the second period; and $8 million for the third period and a minimum of nine wells and 6,000 km² of seismic surveys. Esso-Cadipsa, which came in second, bid $41.1 million for the first period; $20.1 million for the second period; and $4 million for the third period. It also agreed to drill no less than 11 wells and conduct 4,500 km² of seismic surveys. The Bridas-Total group bid $7 million in expenditures for each of the three periods. The Shell group won because of the way in which the expenditures were scaled over the three periods. Shell's local partners held 11 percent of the concession while Esso's local partner (Cadipsa) held only 3 percent. However, this was not supposed to have influenced the outcome.

In August 1979 two additional offshore blocks east of Tierra del Fuego were offered for contracts. The bid for both blocks was won by an Esso consortium which included three domestic firms with a total of 10 percent interest. Late in 1979 and in 1980 tenders were issued for several onshore blocks plus additional offshore blocks. These contracts were bid on the basis of a variable K rather than on the basis of the level of expenditures and number of wells to be drilled. Only one risk contract, that with Union Oil, was signed during 1981.

Early in 1981 the Shell Group announced an offshore oil discovery off Tierra del Fuego only a few months after drilling was initiated. This was the first offshore discovery and the first on a risk contract. Esso drilled its first well off Tierra del Fuego in November 1980. Because of unfavorable weather conditions in the area, it is estimated that at least five years will be required to develop commercial output.[7] The Bridas-Total consortium is also reported to have made oil discoveries.[8]

Criticism of the Risk Contracts

Interest on the part of U.S. and other international oil companies in bidding on Argentina's risk contracts has been relatively low and mainly confined to companies that have operated in Argentina for many years. The lack of interest stems mainly from the nature of the risk contracts, several criticisms of which were pointed out to the author by petroleum company officials. One has to do with the complex formula for determining the price, which involves a number of variables, such as the peso-dollar exchange rate and Argentine wholesale prices, that do not move in relation to one another in a predictable way. A serious difficulty with the formula is that the K factor (which mainly determines the price) declines sharply with production within a given block rather than on a field basis.[9] This means that it might not be profitable to undertake exploration in areas within the block which may be less promising than the prime area. In other words, risk expenditures for expanded production in marginal areas are discouraged by the formula.

A second criticism of the risk contracts is that onshore exploration areas, 80 to 100 km², are regarded as much too small for carrying out modern exploration techniques. The offshore areas are much larger, running to 6,000 km² for each block.

A third criticism is that if the wells should produce mainly natural gas, there is no assured

[7]*Petroleum Economist,* March 3, 1981, p. 121

[8]"YPF Offshore Production Claim Meets Doubts," *Oil and Gas Journal,* May 9, 1983, pp. 78–79.

[9]In the case of the Magellan East bid, the formula was amended to permit application of the K factor on a *field* basis, thereby partially correcting the difficulty.

market for the product during the period of the contract. Moreover, the companies believe that the offshore areas near Tierra del Fuego may yield largely gas wells and they fear there will not be enough pipeline capacity to market the gas. In addition, there would be competition from the abundant and cheap onshore gas.

The areas available for contracts are determined by YPF rather than by the secretary of state for energy, which was the case under an earlier law. YPF is a competitor with the private companies in that it also conducts exploration and development, and it has been accused of keeping the more promising onshore areas for its own development. Private company representatives also point out that they have no recourse to the Argentine government in dealing with a variety of administrative rulings by YPF officials. The only recourse would be to go to arbitration under an arrangement whereby the third arbitrator would be appointed by the president of the Supreme Court, but such an action would be politically dangerous and would not be desirable except in the case of a major breach of contract. As the president of one affiliate of a large international company stated, a minor YPF official could issue a ruling that would cost his company millions of dollars.

The contracts are also faulted by the fact that, until domestic needs are satisfied, the petroleum must be sold to YPF rather than on the international market. It seems unlikely that Argentina will ever be a significant petroleum exporter.

Prospective investors are concerned with exploration risks in relation to potential rewards. A number of private Argentine geologists do not believe there are large reservoirs on the continental shelf. Finding low-yielding wells would not justify the high cost of offshore exploration and development.

Some prospective investors are undoubtedly concerned regarding the general political and economic climate in Argentina. Argentina has a long history of nullifying contracts made by previous administrations. Although the government has made compensation in such cases in the past, the companies inevitably suffered losses. A return to civilian government in Argentina may very well bring into power a highly na-

tionalistic group, such as the Peronistas, who may nullify the contracts as they have in the past.[10] Nor do current economic conditions with an inflation rate in excess of 100 percent inspire confidence in investors. In recent years the rate of depreciation of the peso in terms of the dollar has tended to lag well behind the rate of increase in Argentine prices. This makes for very high costs for local wages and materials in terms of the dollar.

Under risk contracts, companies pay Argentine federal government taxes on net income so they do not anticipate any problem in offsetting these taxes against U.S. corporate tax liabilities. However, a problem may arise from the existence of provincial taxes on *gross* revenue ranging from 1.5 to 2.0 percent. The legality of this tax as applied to risk contracts has been under discussion in Argentina, but it could present a problem of creditability against U.S. tax obligations when production begins under the risk contracts.

Brazil

Background

Brazil's petroleum industry was nationalized before oil was discovered in any significant amount. Petroleum production was reserved for the state by Decree Law 2004 of 1953, and Petroleo Brasileiro, S.A. (Petrobras) was established to serve as the state monopoly over exploration, production, and refining of petroleum. Since then Petrobras has discovered and exploited all of the oil fields now producing in Brazil. However, several major international petroleum companies, including Texaco, Shell, Exxon, Gulf, and Mobil, operate at the retail level.

Brazil's petroleum production was only 2,500 bpd when Petrobras assumed control. Production reached a peak of 175,000 bpd in 1969, and then dropped back to about 170,000 bpd, which satisfied about 20 percent of the country's demand for petroleum in the mid-1970s. In 1981

[10]It is reported that prospective contractors were being asked in early 1983 to accept a condition permitting the civilian government scheduled to take office in 1985 to renegotiate any new agreements. "YPF Offshore," p. 80.

Petrobras increased its output to 215,000 bpd, but with domestic consumption estimated at 1 million bpd, production covered less than 22 percent of requirements.

Following the oil price increases of 1973–74, Brazil's import bill rose to over $3 billion per year and put a severe strain on the country's balance of payments. The popular ideological bias against foreign petroleum companies was so great that even the military government that assumed power in 1964 was reluctant to permit foreign petroleum company operations in Brazil for more than eleven years. Finally, in October 1975 President Ernesto Geisel announced that foreign companies would be invited to explore in Brazil on the basis of a "service contract with a risk clause."

Summary of Model Contract Used in First Three Rounds of Risk-Contract Tenders

The first tenders for the risk contracts were issued in April 1976. According to the model contract,[11] each bidder was required to buy a set of surveys of the 10 prospective areas at a cost of $400,000. Upon winning a bid, the foreign contractor would pay all exploration, evaluation, and development costs, and would be reimbursed for these investments following commercial discoveries in amounts and over a period of time to be determined in part according the bidding. Petrobras retained exclusive ownership of all petroleum and other resources found and would pay for discoveries in money rather than oil. Following exploration and development, Petrobras would assume control of production. Expenditures for exploration and evaluation are repayable without interest; capital expenditures for development are repayable at interest (maximum interest, prime rate plus 1 percent); and the contractor is to receive a percentage of revenue from production valued at world prices. (See the appendix for an outline of conditions governing the April 15, 1976 tender.)

[11]The Brazilian government has forbidden the publication any risk contracts with the oil companies and has not en made public the texts of model contracts issued from ne to time.

Payments to the contractor from commercial production of a field may not exceed the net income of Petrobras from that field during the previous quarter. Whenever such net income is smaller than the payments due, the balance in favor of the contractor is deferred and added to the subsequent payments. The market price in U.S. dollars of crude oil in the fields discovered and developed by the contractor shall be equivalent to the current sales price of crude oil in the international market when sold in freely bargained arms-length, long-term transactions, with due allowance made for quantity, quality, and locational differences. If the contractor believes that the price is not representative of the international market price, the contractor may notify Petrobras and, if the parties are unable to agree, the matter is to be determined by arbitration.

According to the model contract, bidding variables in the first round (1976) were to include (a) a cash bonus (not recoverable as exploration cost); (b) the exploration program and minimum expenditure; (c) number of wells to be drilled; (d) exploration and development investment repayment period and interest rate on development investment; and (e) the contractor's fee as a percent of gross revenue. So far as I am aware, cash bonuses have been modest—on the order of $500,000—and some bids have not included a cash bonus. The major element in bidding has been the percentage of the revenue paid to the contractor. This bid percentage differs with the level of output, rising as output increases.

Petroleum risk contractors are subject to a flat 25 percent tax on net profits and on interest paid on the development expenditures which they have undertaken. This is considerably less than the normal Brazilian corporate income and withholding tax. Petrobras agrees to absorb any increase in the current tax imposed on the remuneration or on the interest, as well as the establishment of new taxes, "provided they are not applicable on a general basis."

Response of Companies to Tenders

Response to the initial terms of the contracts on the part of oil companies was not enthusiastic. On the one onshore and nine offshore areas offered in the first round, only five bids were received from seven companies or consortia. By April 1977, British Petroleum (BP), Elf Aquitaine (AGIP), the Shell Group (Shell International, Pecten Brazil and Enserch), and Exxon had signed contracts. The Shell Group's contract covered a double tract totaling 5,000 km² off the mouth of the Amazon; the Elf/AGIP contract is in the same general area; and Exxon's tract was off Rio de Janeiro. A fifth bidder, Texaco, withdrew from bidding and the area in which it was interested went to Elf/AGIP.

In the second round of tenders for contracts announced in May 1977, bids were sought on 25 offshore blocks, including blocks in the mouth of the Amazon, in the Santos Basin, in Rio Doce, and southeast of Rio Grande. In order to attract more companies, the price of the geological information was reduced to $250,000 and the size of the areas offered was enlarged. Reportedly there were 28 bids, but there were no bids on a number of the blocks offered.

In the third round of bidding announced in 1978, the tenders issued covered onshore as well as offshore tracts. Three blocks in the Amazon River basin southwest of Manaus were awarded to the Shell Group; Marathon was awarded an offshore area off Amapa; and another tract went to a group comprising Brazilian subsidiaries of Chevron, Union Oil, Ocean Drilling and Exploration, and Cities Service. Esso received two tracts, one off the mouth of the Amazon River and one off Bahia. All companies that negotiated agreements in the third round had been awarded tracts during the first two rounds, either alone or in partnership. However, the contracts negotiated covered only a few of the 21 onshore and 21 offshore blocks that were opened for bidding. The combination of poor results of drilling by the private companies plus the unfavorable features of the contracts themselves resulted in relatively little interest on the part of international petroleum companies.

Although the contracts have not been officially made public, the third round Marathon contract of July 1979 was published unofficially in a provincial Brazilian newspaper. According to this contract, Marathon will obtain 28 to 35 percent of any of the oil it discovers for fifteen years. The company will also be reimbursed for its investment with interest (at the Bank of America, prime rate plus 1 percent), and the payments are to be made in dollars with a possibility of payment in crude up to 45 percent of the total. The percentge of oil produced in each commercial field to be paid to Marathon is 35 percent for the first 600,000 barrels during any one quarter; 31 percent for a quarterly output of between 600,000 and 1,200,000; and 28 percent for a quarterly output in excess of 1,200,000. Payments made in dollars are to be based on the market price of crude produced in each field discovered and developed by the contractor. Total payments to Marathon cannot exceed 80 percent of the value of production from the field discovered. Marathon's nonrecoverable cash bonus was $500,000.[12]

Criticism of the Initial Contracts

There were several criticisms of the initial risk contracts, some of which Petrobras and the Brazilian government sought to correct when the fourth round of tenders was announced in 1979. The principal criticisms and requests for changes reportedly made by the companies were as follows:

1. The companies wanted to be paid their portion of the output in petroleum rather than dollars, plus have the freedom to export petroleum to their foreign markets.

2. The companies did not want to be committed to a minimum investment of up to $15 million per block after they had run seismic surveys which indicated poor drilling results.

3. The companies wanted technical information available for an entire basin rather than simply for the blocks on which they were making bids.

4. The companies wanted to operate the oil fields they discovered rather than turn them over to Petrobras following development.

[12]The Brazilian text of the Marathon contract was published in the newspaper, *Diario de Povo*, July 2, 1980.

5. The companies wanted provisions for marketing natural gas and for recovering the full costs of discovering and developing a gas field.

6. The companies stated that Petrobras takes the most attractive areas for its own operations, allocating to the companies only the less attractive areas for exploration.

7. The companies wanted international arbitration of disputes instead of the national arbitration procedures set forth in the contracts.

A further criticism of the risk contracts, which may or may not have been made by private petroleum company officials, is that the formula for remuneration to the contractor, which provides a lower percentage of output to be paid to the contractor with higher levels of output, may work against efficient exploration and development of an area. This is because wells drilled in marginal locations will have a lower probability of success, but if they do succeed, the returns from the investment in the marginal locations will be lower as a consequence of the increased output. There should be some provision in the contracts for higher terms on marginal investments.

The 1980 Model Contract

The Brazilian government evidently took seriously the companies' criticisms of the risk contracts employed in the first three rounds since they liberalized considerably the conditions for the fourth round of bidding. The changes made in the 1980 model contract and in Brazil's petroleum policy generally included the following:[13]

1. Eighty-five percent of Brazil's sedimentary basins are to be opened for exploration by foreign and domestic private companies. Of the total sedimentary basin area of 5.1 million km²—

including offshore areas down to a depth of 2,000 meters—Petrobras reserved 760,000 km² for itself; another 330,000 km² was already contracted to private companies; the remaining 4.1 million km² would be offered to private companies over the next three years. Of the new areas offered, 1.1 million km² would be offshore and 3.0 million km² onshore.

2. Complete basins would be offered for bidding rather than breaking them up into blocks.

3. For a fee, interested companies could obtain data on complete basins rather than on individual blocks.

4. Payments to companies discovering petroleum would be made in oil rather than the dollar equivalent, at the option of the contractor, but subject to certain conditions.

5. The contractors were permitted to participate in certain decisions in the production and marketing stages.

6. Contractors were given a "drilling option" in place of the minimum investment requirement, which included drilling one or two wildcats regardless of the outcome of the geological surveys. Under the drilling option, a contract company is obligated to invest only in seismic surveys, and if there are indications of good prospects, it has an option to drill wildcats. If seismic profiles are negative, the block may be turned back without further outlays. If the contractor decides to drill, he must agree to the minimum investment.

7. The provisions that apply to petroleum apply also to natural gas that can be economically developed and produced for supplying existing markets, or if a market can be developed within five years of discovery. If a market cannot be developed within this time, the contractor is not entitled to any remuneration or reimbursement with respect to the gas discovery. The price of natural gas is to be a function of the price established for the market "after joint negotiations between the potential consumers and Petrobras with assistance of the contractor," provided that due account shall be taken of (a) the quantity and quality of the natural gas; (b) the price of natural gas from other Brazilian sources; (c) the price at which supplies of natural gas imported

---

[13]A copy of the model service contract issued in late 1980 was made available to the author by an official of the Brazilian government with the request that it not be published, but I was given permission to summarize it.

into the onshore region in the vicinity of the service area are being made; and (d) the international price of competing or alternative fuels.

Bidding under Fourth, Fifth, and Sixth Rounds

In the fourth round, 123 blocks were offered, including 99 blocks onshore totaling 1.1 million km², and 24 blocks offshore totaling 67,840 km². The fourth round drew 34 bids, more than any of the previous rounds. Bids came from 14 companies which separately or in consortia were seeking contracts for 13 onshore and 21 offshore areas. Four new foreign companies bidding for the first time were Conoco (U.S.), Deminex (Germany), Husky Oil (U.S.), and Brazil Petroleum Exploration (Japan). Seven Brazilian companies also bid. By the end of 1980, a total of 30 risk contracts had been signed in the fourth round out of the 123 blocks offered. Also, by that date 80 risk contracts had been signed since the beginning of the program. Table 9-2 shows the breakdown of companies and blocks signed as of early November 1980.

Under the fifth round of bidding in 1981, 150 blocks were offered covering 460,000 km² of offshore areas in the northern half of the Brazilian continental shelf from Bahia to the French Guyana border. This was followed by the announcement of a sixth round of risk-contract solicitations offering some 700 onshore blocks covering 2 million km².[14] However, there were only two bidders, Esso Inter-America and Anschutz Overseas, and both companies signed contracts covering an area totaling less than 15,000 km².

Results of Risk-Contract Program

By mid-1982, 107 risk contracts had been awarded to some 34 private contractors. More than 50 wildcat wells had been drilled by these contractors with only one discovery reported. Petrobras has made important discoveries of both

[14]*Brasil Energy* (Rio de Janeiro), May 10, 1981, p. 1; see also, "Brazil Opens More Acreage to Bidding," *Oil and Gas Journal*, September 6, 1982, p. 44.

Table 9-2. Foreign Petroleum Companies that Have Negotiated Risk Contracts in Brazil and Location of Blocks as of November 1980

Company	No. of contracts	Location
BP/Citco	6	All offshore Santos
Shell/Pecten/Enserch	2	Onshore Parana
Marathon	1	Offshore Amazon
Elf	4	1 offshore Santos
		1 offshore Amazon
		1 offshore Ceara
		1 onshore Middle Amazon
Esso	12	8 offshore Amazon
		1 onshore Amazon
		2 offshore Santos
		1 offshore Bahia
Pennzoil	2	All offshore Santos
Hispanoil	1	Offshore Amazon
Hispanoil Hudbay	1	Offshore Espirito Santo
Hispanoil Hudbay/ Deminex	2	Offshore Maranhão
Pecten/Shell	5	4 onshore Middle Amazon
		1 offshore Amazon
Pecten/Shell/Marathon	1	Offshore Santos
Pecten/Marathon	1	Offshore Santos
Citco/Union Oil	1	Offshore Amazon
Pecten/Chevron/ Marathon	1	Offshore Santos
Marathon	1	Offshore Amazon
Pecten/Chevron/Union Oil	2	Offshore Bahia
Citco/Chevron/Union Oil/Canam	4	All offshore Maranhão
IPT/CESP	17	All onshore Sao Paulo
Union Oil/Brapex	1	Offshore Alagoas
Husky Oil	1	Offshore Ceara

Source: *Brasil Energy*, vol. 1, no. 19, November 10, 1980, p. 2.

crude and gas in the Campos Basin off the state of Rio de Janeiro and, in addition, has reported some wildcat strikes in regions adjacent to some of the new blocks on which contracts with private companies have been awarded. The first significant discovery by a private contractor was announced in January 1982 by Pecten (Royal Dutch Shell group) in the Camamu Basin off the state of Bahia. It was reported that the test well is capable of producing more than 1,000 bpd of 31° gravity crude. However, the significance of the discovery cannot be assessed until there is further drilling.[15]

Although the liberalization of risk contract terms has substantially increased the interest of private petroleum companies in bidding, thus far bids have been made on only a small proportion of the total blocks offered. Moreover, relatively little interest was shown in the sixth round of bids. Hopefully the Pecten discovery will prove significant and, if it is followed by other discoveries, increased bidding and drilling operations are likely to take place.

Evaluation of Argentine and Brazilian Service Contracts

The Argentine exploitation contract for producing oil from areas or zones on which exploration has been conducted differs from an exploration-exploitation contract only by a reduction in the degree of risk to the contractor since there is no assurance that wells will be productive or what their cost and output will be. Contracts providing for the development of a field on which exploration has taken place stipulate the number of wells to be drilled (or expenditures undertaken) and payment on the basis of a price determined by a formula which includes a number of uncertain variables. The inability of the companies to adjust exploration outlays to the amount warranted by initial exploration results increases risk and impairs operational efficiency. Moreover, price bidding has much the same effect on the incentive to produce high-cost/low-volume fields (or undertake high-cost secondary recovery) as royalty bidding.

Since both Argentina and Brazil need to expand their reserves, the risk contracts are better suited to this requirement. Of the two risk-contract models, the Brazilian is perhaps somewhat superior to the Argentine model, although both may be faulted by contract provisions specifying the number of wells to be drilled or the minimum

expenditure. (This disadvantage was only partially relieved by the introduction of the "drilling option" in the 1980 Brazilian risk-contract model.)

The Argentine risk contract employs a complex formula for determining the price to be paid to the contractor for the oil, with the price declining as the volume of production rises. This arrangement works against maximum production and increases risk as a consequence of the variables of the formula. The Brazilian system provides for payment to the contractor for the oil at the world market price, or in oil at the option of the contractor (1980 model contract). However, under the Brazilian system, the contractor's remuneration takes the form of a certain percentage of the oil produced (or gross revenue) at the world price, with the percentage declining with the volume of oil produced. (This percentage is a bid variable.) The Brazilian contracts provide for repayment with interest of exploration and development expenditures of the contractor. Such investment recovery is not provided for in the Argentine contract. Investment recovery tends to reduce risk, although no recovery is provided under the Brazilian system unless a discovery is made and sufficient oil is produced. The Argentine system does provide for the rapid depreciation of equipment in the calculation of taxable income.

Under the Argentine risk-contract system, the contractor retains control of operations during the exploitation period, while under the Brazilian system operations are under the control of the government enterprise, Petrobras, following exploration and development. This is regarded by companies as a serious flaw in the contract, which is only partially removed by a provision in the 1980 model contract that contractors may participate in certain decisions in the production stage and in the decision to undertake commercial production.

Finally, the Argentine tax rate on net income is higher than the Brazilian rate, but the effective rate may depend upon the effects of differing arrangements with respect to depreciation.

Clearly, the interest of foreign companies in bidding on risk contracts has been substantially greater in Brazil than in Argentina, where only

[15]"Pecten Group Drills Hefty Oil Discovery Off Brazil," *Oil and Gas Journal,* January 11, 1982, p. 134; "Brazil Opens More Acreage," *Oil and Gas Journal;* and F. E. Niering, Jr., "Brazil: Steady Rise in Oil Production," *Petroleum Economist,* April 1982, pp. 149–150.

a few companies have negotiated risk contracts. However, Brazil has made much larger areas available for bidding and only a small percentage of these areas have been bid upon. There is also evidence that the high degree of interest on the part of foreign companies in 1980 and 1981 has tended to subside. It would appear that unless important discoveries of large fields are made in Brazil, further interest in exploration will be limited. This interest could undoubtedly greatly expand if the contracts were made more attractive. Also, in both Argentina and Brazil the high rates of inflation and the very serious debt problems provide an unfavorable climate for investment of any kind. High rates of inflation produce uncertainties with respect to costs, while serious balance of payments problems render companies vulnerable to exchange controls that limit or prohibit the repatriation of earnings.

APPENDIX
OUTLINE OF PETROBRAS INITIAL MODEL CONTRACT, APRIL 1976*

BID PROCEDURES

July 15, 1976 submission

Bid on any number of blocks. Three maximum awarded per contractor

Must bid according to established procedures and acceptance of terms of contract. Nonconforming bids not considered.

$300,000 bank guarantee per block. Lose if do not sign contract within 30 days after being invited to do so.

Bid variables include:

Cash bonus—not recoverable as exploration cost

Exploration program and spend-or-pay minimum

Drilling spud date

Exploration and development investment repayment period and interest rate on development investment

Contractor fee as percent of gross revenue (function of annual production) and period fee applicable

Petrobras selects bidders with which it negotiates

EXPLORATION

Area:

0.6 million—1.2 million acres offshore and 3.5 million acres onshore—maximum of 3 areas per contractor

Term:

3 years offshore and 5 years onshore, starting with date contract signed (e.g., July 15, 1976 + negotiation period + invitation to sign + 30 days)

Commerciality must be established by end of period or contract terminates

Relinquishment:

50% at end of ⅓ exploration period

25% at end of ⅔ of exploration period

Bonus on Signing:

Bid variable—not recoverable as exploration cost

Exploration Program:

Bid variable—work program with spend-or-pay minimum (minimum guaranteed by bank)

Programs and budgets approved by Petrobras

Contractor subject to current importation laws—not fixed for life of contract. Duties payable by contractor—recoverable if commercial field found. $20,000 aggregate per month limit for personnel salaries and costs.

Complete Petrobras operations control

Drilling Commitment:

Bid variable—timing of first well

Determination of Commerciality:

Subject to absolute discretion of Petrobras, on basis of contractor reports, by end of period or contract terminates

Precondition to development period and eventual reimbursement of expenses

Formula vague and subjective

Natural Gas:

No contractor rights to remuneration, reimbursement or development. Petrobras may decide to agree on contractor development of field.

DEVELOPMENT

Program:

Period, program and budgets proposed by contractor with Petrobras right to require changes with no apparent contractor recourse

Development Period After Which Contract Terminates:

Time fixed by contractor with Petrobras approval

Period ends when development completed and installations accepted and received by Petrobras

Exploration and Development Investments:

For contractor account

*Source: Confidential.

Assets become property of Petrobras on importation into Brazil

Producing Operator:

Petrobras, without contractor participation—but contractor has obligation to give 6 month's technical advice

Contractor Payments:

1. Exploration and Development Investments

    Loan concept with repayment term a bid variable

    Exploration investment—no interest

    Development investment—maximum interest prime + 1%—(bid variable)

    Interest starts when development complete and payable on outstanding balance

2. Continuing Fee

    Percent of revenues as a function of annual production—(bid variable)

    Term of fee bid variable

3. Payments delayed one year after production and limited to Petrobras net income from field (defined as gross income less direct costs of production, collection, storage and transportation, overhead, Brazilian severance taxes)

Contractor Oil Purchase Entitlement:

Contractor option to purchase (for payback period) at market price established by Petrobras, crude oil from field discovered equal to value of amounts paid back to contractor

Option suspendable by Petrobras if "national supply of petroleum comes to a crisis"

Arbitration:

Contract so one-sided in favor of Petrobras, arbitrators would be hard pressed to find for contractor

Brazilian laws apply

Force Majeure:

No suspension of time periods in contract because of Force Majeure

Foreign Exchange, Taxes and Duties:

Generally applicable laws apply—

No currency denominated in contract for any purpose (crude pricing, accounting, interest, reimbursement or remuneration payments, etc.) and no devaluation protection on deferred local currency payments

No protection against discriminatory exchange rates (Brazilian law can set at will)

No provision for remittance of exploration/development investment reimbursement as capital repatriation under existing law (which reduces profit remittance base)

Remuneration and interest on reimbursement is subject to existing profit remittance law

Income and profits taxes, including a supplementary income tax on remitted profits, result in total effective rates as follows:

Profit Remittances: % of Reg. Cap.	Incremental Supplementary Tax Rate — %	Combined Effective Tax Rate on Profits Remitted—%
0 - 12	---	50
12 - 15	40	50 - 54
15 - 25	50	54 - 62
25 - 100	60	62 - 76
100	60	76 - 80

(The profits tax later changed to a flat rate of 25 percent)

When Brazil has foreign exchange problems, law permits:

a) Limitation of annual profit remittances to 10% registered capital

b) Suspension of capital repatriation

c) 50% surcharge on "financial transfers"

Export proceeds cannot be retained abroad

Duties payable but recoverable only if there is a discovery

10

The Joint Venture and the Modern Concession Agreement

The joint-venture agreement involving a private company and a GOE provides for sharing the risk and capital expenditures, usually for the period during the development and production stages following a successful exploration financed solely by the private company. Although the private company usually has complete control over operations, control over the budget and other policy matters is in the hands of a joint committee made up of representatives of the GOE and the private company. The joint venture, which normally provides for 50-50 equity ownership, may be distinguished from a government participation in which the government has an option to acquire a minority equity (10 to 20 percent) at book value (or sometimes without cost) following the exploration or development period. The joint venture has been used by Indonesia and Malaysia in combination with a production-sharing contract, and a joint venture with majority government equity ownership has been used by Angola, Bahrain, and Cameroon. The Colombia contracts of association are perhaps the best modern example of the joint-venture agreement in petroleum.

Those petroleum agreements that do not fit

the conditions for production-sharing, service, or joint-venture contracts may be regarded as modern concession agreements. Although they differ substantially from country to country, the basic feature is that government revenue is derived mainly from royalties and net income or profit taxes rather than from some type of product or revenue sharing. The concession form is employed by several oil-producing countries, including Brunei, Trinidad and Tobago, Tunisia, Turkey, and Zaire. Provision for government minority participation in a concession company is quite common for this type of arrangement.

The Papua New Guinea (PNG) petroleum agreement is an example of the modern concession agreement. The PNG model agreement is of special interest because it employs the resource rent tax as a form of progressive income tax. Until 1969, Colombia employed concession contracts exclusively and, although new concession contracts are no longer granted, the bulk of the oil currently produced in Colombia comes from fields covered by concession contracts. However, the Colombian concession contracts differ from most modern concession contracts in that all oil produced must be sold to the government at a price determined by the government, which bears little relationship to the current world price.

The Colombian Joint-Venture Contracts

Early History of Colombian Petroleum Contracts

Private concessions for petroleum exploration were granted in Colombia as early as 1905, and the existence of oil in commercial quantities proved by 1916. Commercial crude production was initiated in the early 1920s by the Tropical Oil Company (a subsidiary of Standard Oil Company of N.J.) which held the "Concession de Mares" and found oil in the Middle Magdalena Valley. Tropical Oil's original concession contract expired in 1951 and the assets were taken over by the Colombian state enterprise, Empresa Colombiana de Petroleos (Ecopetrol). Standard Oil (now Exxon) formed a new subsidiary, International Petroleum Colombia Ltd. (Intercol), which continues to operate in Colombia. Meanwhile other petroleum companies, including Texaco, Cities Service, and Gulf, obtained concessions and began producing oil and gas.

Colombia's crude production reached a peak in 1970 of about 220,000 barrels per day (bpd), but has declined since then as reserves in the traditional producing areas, such as the Magdalena Valley, ran down. The discovery in 1963 by Texaco and Gulf of the Orito field in the Putumayo region close to the border of Ecuador was expected to achieve production of 150,000 bpd, but the fields in this region have proved to be disappointing and production from the Orito field in 1979 was only 22,500 bpd, in contrast to 80,000 bpd in 1969. By 1978 domestic

production had declined to 130,000 bpd, with a further decline to 125,800 bpd in 1980. Colombia shifted from being a net exporter of 80,000 bpd in 1970 to a net importer in 1976, and in 1980 net petroleum imports supplied over 20 percent of domestic consumption. Recent discoveries are expected to raise Colombia's output to over 150,000 bpd by 1985.[2]

The Concession Contracts

Until 1969 the exploration and development of petroleum and gas by the private sector was entirely under concession contracts. Royalties varied between 11.5 and 14.5 percent and, in addition, there were income taxes on net profits. The period of exploration ranged from three to four years and exploitation ranged from thirty years in the central zone to forty years in the eastern zone, with a ten-year extension permitted in both cases. As of December 1978, some 30 concession contracts were still in existence. Among the companies holding concessions were Intercol (Exxon), Texaco, Chevron, Shell, Cities Service, Houston Oil Colombiana, and Elf Aquitaine. Most of Colombia's oil production currently comes from old oil fields covered by concession contracts, but by 1990 the bulk of the output is expected to come from new fields discovered since 1979. In 1978, 43.6 percent of Colombia's petroleum output was produced by Ecopetrol and 56.4 percent by private companies, most of which operated under concession contracts.

Under both the concession contracts and the contracts of association, all petroleum must be sold to Ecopetrol at prices established by the Ministry of Mines and Energy. In 1979 the base price for crude produced under concession contracts was only $1.50 per barrel, but companies could negotiate somewhat higher prices as an incentive for increasing production, based on increased output in excess of a declamation curve which described the expected life of the field.

In 1979 Texaco sold all its interests in the Putumayo oil-producing region, including its stake in the Trans-Andean Pipeline, to Ecopetrol. Gulf Oil withdrew from the partnership with Texaco in 1973. Texaco had a contract of association with Ecopetrol.

[2]F. E. Niering, Jr., "Colombia: Upturn in Oil Output and Drilling," *Petroleum Economist* (February 1982) pp. 50–52.

In early 1980 the price for incremental oil varied from $1.75 to $8.50 per barrel. The low base prices for crude established by the government and the delays in negotiating higher prices were largely responsible for the decline in production during the 1970s. One of the reasons Texaco sold its 50 percent equity in the oil operations of the Putumayo area in 1979 was that it received only $1.70 per barrel for the crude oil produced.[3]

In May 1980 the government published Resolution 058 which increased the base price for production from existing fields, established a mechanism for annual price rises, and a formula to determine prices for future incremental output from older fields. The annual price increase is a function of the Colombian price index and the rate of devaluation of the Colombian peso. Most company officials believe that output from the old concession fields could be increased by 20 percent or more if the price were raised in line with world levels. In 1981 oil prices set by the government rose by 13.1 percent from 1980 levels.

Contracts of Association

Law 20 of 1969 stated that "The government will be entitled to declare any petroleum area of the country as a national reserve and assign it to Ecopetrol, either for direct exploitation or for development with either public or private capital." In addition, the law authorized Ecopetrol to negotiate contracts of association with private companies in which Ecopetrol would benefit both through royalties and through a share in production. Concession contracts continued to be negotiated after 1969, but since 1974 only contracts of association have been negotiated.

The contracts of association provide that all of the investment in development and the operating expenses are shared 50-50 by the associated company and Ecopetrol. Ecopetrol reimburses the contractor for 50 percent of the exploration costs of discovery wells. Production is distributed on a 50-50 basis but, after paying a royalty (which may vary from 16 to 30 percent but is normally 20 percent), the contractor receives 40 percent of the output. The contractor pays the royalty on *all* of the output. In addition, the private company must pay a corporate tax of 40 percent of net income. Beginning in 1976, new contracts of association provide for payment of the contractor's share of oil at the world price.

Companies that have negotiated contracts of association include Texaco, Amoco, Elf Aquitaine, Cities Service, Mobil, Pennzoil, Phillips, Gulf, Sunray Colombia Oil Company, Houston Oil and Minerals, Arco, Exxon, Occidental and Chevron, some of which are in joint ventures with other companies. Since 1974 some 50 contracts of association involving more than 25 companies, either singly or as members of consortia, have been signed through 1981.[4] However, some of these contracts have been terminated and only a few of the associated companies have found oil in the contract areas. In 1980 and 1981 Exxon and an Elf Aquitaine group made important discoveries in the Eastern Plains or Llanos Orientales; these discoveries have substantially increased the interest of petroleum companies in negotiating contracts in this region.[5] In 1980, 20 new contracts of association were signed and additional contracts were reported to have been signed in 1981.

Model Contract of Association of November 1976

Although contracts of association have been negotiated since 1970, most of them have been terminated either because the contractors failed to find oil or because the price established by the government for valuation of the contractor's share of oil was too low to make the operation profitable. The new model contract of association approved in November 1976 provided for valuation of the contractor's share of oil produced at the international c.i.f. price set at Cartegena, Colombia for similar crudes. In addition, the contractor was free to export his share of

[3]"Texaco Sells Colombian Interests," *Petroleum Economist* (December 1979) p. 520.

[4]Niering, "Colombia: Upturn," p. 50.
[5]Ibid.

the oil subject to limitations by the government. However, in 1978 the government decreed that all petroleum produced by the contractors must be sold to Ecopetrol at the world price, and Ecopetrol is the sole exporter of petroleum.

The contracts of association negotiated since November 1976 have not departed substantially from the 1976 model contract. Where significant changes have occurred, I have followed the language of actual contracts in my possession rather than that of the model.

NEGOTIATION OF CONTRACTS. Ecopetrol has the right to select both the areas available for contracts of association and the prospective contractors. However, prospective contractors may suggest particular areas for contracts with Ecopetrol. There is no formal bidding, although presumably this could occur informally if more than one firm were interested in a particular tract. There is no legal limit to the size of the contract areas, and firms or consortia may have more than one contract of association, each covering a particular tract. Contract areas have ranged from 7,000 to over 1 million hectares. In early 1980 Provincia Petroleum obtained contracts for three areas, each of which totaled about 1 million hectares in the area of Los Llanos Orientales; while Chevron Petroleum's recent contract area covered about 150,000 hectares. The minimum number of exploratory wells which the contractor is required to drill and the minimum expenditure for exploration, including seismic surveys, are negotiated between the contractor and Ecopetrol. An option to terminate the contract after the seismic survey appears in some of the contracts negotiated after 1976. This differs from the model contract of association which requires drilling of at least one exploratory well before the contractor can terminate the contract. Contractors are required to pay Ecopetrol the amount of any unexpended funds committed in the contract. The work program, including the number of years from the effective date of the agreement (usually six years) that the contractor has to complete the exploration program, is also negotiated. Once a commercial discovery is made, an executive committee composed of a representative from Ecopetrol and one from the con-

tractor is established; they decide on the development program and the budget.

EXPLORATION. During the initial exploration period, the contractor agrees to conduct seismic exploration programs in the contract area according to an agreed-upon number of kilometers of lines or to spend a certain amount of money. Upon completion of this program, the contractor may elect to proceed with exploratory drilling or terminate the contract. After the second year of the contract, the contractor has the option of relinquishing the contract area provided the minimum amounts agreed to be spent for the seismic survey and exploratory drilling have been met. If the contractor decides to undertake additional drilling beyond the first exploratory well, he is obligated to drill a minimum number of wells within a six-year period. This minimum has ranged from four to fifteen in the contracts negotiated thus far.

Upon termination of the initial exploration period (three years), an extension for an additional three years may be requested by the contractor. If a commercial field has been discovered in the contract area, the area must be reduced to 50 percent of the original area at the end of the exploration period; two years later the area must be reduced to 25 percent of the area originally contracted; and two years thereafter the area must be reduced to the area of field or fields under production or development, plus a 5-km-wide reservoir zone surrounding each field. The contractor may determine the areas to be returned to Ecopetrol in blocks of no less than 5,000 hectares each, unless the contractor demonstrates this is not possible. The contractor is not obligated to return areas that are under development or production.

EXPLOITATION. If the existence of a commercial field is determined on the basis of a sufficient number of exploratory wells, the contractor shall advise Ecopetrol, which may accept or reject the existence of the commercial field. If Ecopetrol accepts, it reimburses the contractor for 50 percent of the costs of exploratory drilling and of completing and operating those wells which have proved commercially productive and are placed in production by the operator. Other exploration costs, including seismic surveys, dry wells, etc., are borne 100 percent by the contractor. The payments are made to the contractor from Ecopetrol's participation in the first production of such wells. If Ecopetrol does not accept the existence of a commercial field, it may indicate to the contractor that additional operations may be necessary to prove the existence of a commercial field and the exploration period may be extended by an agreed-upon period with an expenditure not to exceed an agreed-upon amount. If Ecopetrol does not accept the existence of a commercial field after the completion of the additional exploration, the contractor has the right to carry out exploitation of the field and to reimburse itself for 200 percent of the total expenses for drilling the productive wells. Thereafter, the field becomes the property of the joint account free of any further exploration or development costs.

The contractor has full control of all operations in the exploitation of the field. The contractor is required to present annual budgets for final approval by an executive committee made up of a representative of Ecopetrol and of the contractor. The contractor determines the maximum productive efficiency rate (MER) for each commercial field and prepares the spacing and programming of the drilling of wells on the basis of economy and efficiency, subject to approval of the executive committee.

THE JOINT ACCOUNT.   At the time that the parties accept the existence of a commercial field, all expenses, payments, investment costs, and obligations incurred for the performance of the operations and the investments made by the contractor before and after the recognition of a commercial field, including drilling and completing producing wells within the field, are charged to a joint account into which both parties make payments on a 50-50 basis. In the case of delinquency on the part of one party, the other party is entitled to interest equal to one-and one-half times the commercial interest rate.

ROYALTIES   AND   DISTRIBUTION   OF OUTPUT.   The contractor delivers to Ecopetrol a royalty (normally) equal to 20 percent of the *total* production of petroleum from the area, and 20 percent of the net gas production.

After deducting royalties and costs as indicated above, the remainder of the petroleum and gas produced from the field is shared by Ecopetrol and the contractor on a 50-50 basis.

OPERATIONS AT THE RISK OF ONE OF THE PARTIES.   If at any time one of the parties wishes to drill an exploratory well not approved under the operating program by the executive commitee, the contractor, on request of the participating party, may drill the well for that party's account and risk. If the well is completed, production from the well (after deducting the royalty) becomes the property of the participating party until such time as the net value of the production (after deducting production costs, etc.) is equal to 200 percent of the cost of drilling and completing the well; thereafter, it becomes the property of the joint account.

MAXIMUM DURATION.   The contracts have a duration of no more than twenty-eight years divided as follows: up to six years as exploration period and twenty-two years as exploitation period, counted from the date of termination of the exploration period for each field.

PRICE FOR CRUDE OIL.   The price for crude oil which is paid to the contractor by Ecopetrol, in lieu of the receipt of oil, is the c.i.f. Cartegena price of similar crudes, determined by the average sales price of crude f.o.b. three Colombian ports that regularly supply the east coast market of the United States.

TAXATION.   Where foreign companies operate as branches, it is generally possible to avoid the Colombian 20 percent dividend remittance tax, so they pay only the 40 percent corporate tax. Interest is deductible for calculating taxable income if the external loan is registered and approved by the exchange control authorities. Although depreciation schedules are rather liberal, Colombian tax law provides that the base for depreciation is peso expenditures with no adjustment for inflation or changes in the exchange rate. This substantially reduces the effective annual amount of depreciation.

ARBITRATION. International arbitration is forbidden under Colombian law. If a disagreement between the parties relating to the interpretation and performance of the contract cannot be settled amicably, the dispute may be referred to final decision by experts appointed as follows: one by each party and a third appointed by mutual agreement; if such agreement cannot be reached, the third expert shall be designated by the board of directors of the Sociedad Colombiana de Ingenieros of Bogota. Any difference of an accounting nature shall be settled by a decision of expert certified public accountants, with the third expert designated by the Central Board of Accountants of Bogota.

Evaluation of Contracts of Association

The Colombian contracts of association are among the most liberal petroleum contracts in the world from the standpoint of the private contractor. Although exploration costs must be borne 100 percent by the contractor, except for those exploratory wells that turn out to be commercially productive, development and operating costs are divided equally between Ecopetrol and the contractor. If we assume a world price of $30 per barrel, an exploration cost not related to producing wells of $2 per barrel, a development and operating cost of $2 per barrel (shared with Ecopetrol), a 20 percent royalty, and a 40 percent tax on net profits, the contractor has an after-tax return of $6 per barrel on which he would probably not have to pay much, if any, U.S. corporate tax. This would provide a net return (after development and operating costs) to the contractor of 12 percent of gross revenue. (This net return does not include exploration costs.)

One criticism of the contracts of association which was expressed by an American oil company official is that Ecopetrol wants to keep certain areas for itself, particularly those on which it has done exploration. For example, one company wanted an area near the Ecuadorian border for a contract of association, but was unable to

obtain it from Ecopetrol. Another criticism is that the contractor has no assurance of the price he will receive for natural gas, or whether there will be a market for the gas. However, this is a problem common to many areas of the world.

Perhaps the major deficiency of the Colombian contracts of association is the 20 percent royalty on total output, or 40 percent on the company's share of total revenue. This high effective royalty could discourage the production of marginal fields and secondary recovery of existing wells. It could also render high-cost/low-volume fields uneconomic from the standpoint of the private company, and thereby reduce the expected IRR—especially since Colombian fields have a relatively low volume of reserves. Moreover, two of the areas under exploration in Colombia are known to have heavy oil, which tends to be expensive to produce.[6]

Another difficulty with the Colombian contracts of association is that the private contractor must undertake all the exploration risk and receives only a half interest in the production. This arrangement, together with the requirement to pay the royalty on total output (both the contractor's and the government's share), greatly reduces the expected NPV or IRR from an exploration investment. Moreover, since the fiscal system is regressive in the sense of taking a higher percentage of net revenue from lower cost fields, the risk-corrected IRR is likely to be substantially higher than in the case of a fiscal system that permits profitable operation of low-quality fields.[7]

The Papua New Guinea (PNG) Concession Contract

The PNG Petroleum Act of 1977 established a relatively liberal regime for private petroleum investment and for this reason, plus PNG's unique tax law as applied to petroleum, this country's

[6]"Colombia," *International Petroleum Encyclopedia 1982* (Tulsa, Okla., PennWell Publishing Co., 1982), vol. 15, pp. 166–167.

[7]The high royalty and the production-sharing aspects of the contracts make for regressivity. This is only partially compensated for by sharing development and operating costs, but such sharing probably does not offset the private contractor's burden of having to pay royalty on total output.

agreements are worth reviewing. Ten prospecting licenses have been issued in recent years, but as of the time of writing no commercial discoveries had been made. The major companies that have been partners in private joint ventures that have obtained prospecting licenses from the PNG government include Canada Superior, Esso East (now withdrawn), British Petroleum, Mobil, Gulf, Newmont Oil, St. Joe Petroleum, General Crude, and Australasian. There was little exploration activity in 1980 and 1981, and activity is likely to be low until a commercial discovery is made.[8]

PNG does not have a state petroleum enterprise, and according to PNG's Finance Minister, Barry Holloway, the government is not interested in majority shareholding.[9] The model petroleum prospecting license agreement issued by the government in 1980 gives the state an option to acquire up to a 22.5 percent equity interest in the venture, provided notice is given by the government within four months after a development license is issued. However, the government may negotiate for up to a maximum of 30 percent equity. The oil companies are compensated for the state's share of the venture by payment of the state's proportionate share of the allowable exploration and other expenditures.

In order to carry out exploration, oil companies must first obtain a prospecting license which is awarded under a system of competitive bidding for both onshore and offshore blocks. Bids are evaluated on the basis of the amount of exploration work the company promises to perform, following which the companies negotiate a prospecting agreement covering specified blocks. The prospecting license also covers the terms and conditions under which development of commercial discoveries will take place if there is a commercial discovery. An annual fee of 100 kina[10] per block is levied, with the fee increasing each year to a maximum of 800 kina in the fifth and following years.

If petroleum is discovered, holders of prospecting licenses have the right to apply for a

development license. The application must be accompanied by detailed proposals satisfactory to the state and there is an annual fee for a petroleum development license of 50,000 kina. There is also a royalty of 1.25 percent of the wellhead value of all petroleum produced from the licensed area.

The basic tax rate is 50 percent of net income, with a resource rent tax of 50 percent which cuts in after the investor has achieved an IRR of 27 percent. (The resource rent tax is explained in the appendix to chapter 4.) The oil companies are free to export at world prices their share of oil produced, and they may also be required to market the state's share. The agreement provides protection against new taxes or levies on the export of petroleum and against other taxes on production or income not provided in the agreement. The oil companies are subject to training and localization requirements and also are required to give preference to local supplies when they are available in PNG on competitive terms. Disputes arising under the agreement may be submitted to the jurisdiction of the International Centre for the Settlement of Investment Disputes (ICSID).

Although the contracts are quite attractive, costs of operations in remote areas such as the Fly River Basin (where the indications of hydrocarbons are the most encouraging) are high.

Evaluation of PNG Contracts

The results of the simulations of the operations of the fiscal systems of five countries undertaken by Kemp and Rose (see table 4-4) showed that the PNG system is progressive in the sense that the government's take as a percentage of net income declined for lower quality fields, and that the real IRR for all four hypothetical oil fields is well above the minimum required to attract petroleum companies. Of course, no fiscal system will attract companies to explore where the probability of a discovery is very low. If and when a significant discovery is made in PNG, the liberal fiscal system will undoubtedly attract a number of petroleum companies. The progressivity of the PNG system arises from the

[8]*Petroleum News* (January 1982) p. 35.
[9]*Petroleum News* (January 1981) p. 48.

derived from the income tax and the progressive resource rent tax. Since the royalty is low, only 1.25 percent, it is unlikely to discourage the development of marginal fields.

It should be mentioned that most concession contracts are not as liberal as that of PNG. For example, Tunisia has a 12.5 percent royalty and a net profits tax of 80 percent. Turkey also has a 12.5 percent royalty with net income subject to the general corporate tax regime. Zaire's concession agreements provide for granting 20 percent of the equity shares to the government without compensation, a 12.5 percent royalty, and a 50 percent net profits tax. The granting of the shares to the government has the effect of increasing the nct profits tax to 60 percent.

III

U.S. GOVERNMENT
AND INTERNATIONAL AGENCY
PROGRAMS AND POLICIES

11

U.S. Government Policies and Activities Relating to Petroleum Development in the Non-OPEC LDCs

Introduction

Petroleum exploration and production in foreign areas outside the Middle East are regarded as important for enhancing the security of oil supplies for the United States and other industrial countries and for meeting the energy requirements of oil-importing developing countries. Nevertheless, the U.S. government has done very little to promote petroleum development abroad. There have been a number of congressional hearings on the subject and the General Accounting Office has issued several reports concerning measures for reducing U.S. dependence on Middle East oil by promoting exploration and production in other foreign areas.[1] A 1977 presidential directive established the International Energy Development Program (IEDP) which has operated within the Department of

Energy (DOE). The purpose of the IEDP was to help developing countries meet their energy needs through increased reliance on indigenous sources. The initial program was designed to (1) analyze LDC energy needs, uses, and resources; (2) encourage exploration and development of conventional energy resources; (3) provide research, development, and application of modern energy technologies; and (4) provide training and education and institution development. The activities of the IEDP have been mainly confined to the assessment of the needs and resources in a handful of countries.[2] More recently the DOE, in association with the U.S. Geological Survey, initiated the Foreign Energy Supply Assessment Program to provide reliable information on world-wide petroleum resources and potential by country. Both of these programs were underfunded and understaffed and have done relatively little to promote foreign petroleum development. The Agency for International Development (AID)

[1]Report by the Comptroller General, *The Potential for Diversifying Oil Imports by Accelerating World-Wide Oil Exploration and Production* (Washington, D.C., U.S. General Accounting Office, November 25, 1980) and Report by the Comptroller General, *U.S. Energy Assistance to Developing Countries: Clarification and Coordination Needed* (Washington, D.C., U.S. General Accounting Office, March 28, 1980).

[2]The IEDP program has been terminated. A major rationale for the program was to discourage LDCs from developing nuclear power.

has provided a certain amount of technical assistance for the development of various forms of energy resources in the LDCs.

During both the Republican and Democratic administrations the U.S. government's position has been that, aside from geological surveys and energy assessment, planning and research, petroleum should be a private sector activity. Moreover, any bilateral or multilateral assistance to foreign petroleum exploration and development should be linked with the encouragement of the flow of private equity and debt capital to the petroleum sector. Unlike several other industrial countries that have government petroleum enterprises that operate in foreign areas and, in addition, maintain close relations with and provide assistance to their private petroleum companies, the United States relies almost entirely on private oil companies for its petroleum supplies, both imported and domestically produced; and the government's relationships with private petroleum companies have often been more adversarial in nature than cooperative as far as their overseas operations are concerned.

The remainder of this chapter will be devoted to the activities of the Overseas Private Investment Corporation (OPIC) and to U.S. tax policy with respect to the foreign operations of U.S. petroleum companies.

OPIC Activities in Support of Foreign Petroleum Investment

The principal initiative by OPIC in support of U.S. policy was the adoption of a program of insurance coverage designed to meet the needs of U.S. investors in foreign mineral and energy projects. This program, which was announced in 1977, offered insurance against risks of expropriation, war, internal revolution, and currency inconvertibility for both equity investments and loans for petroleum exploration and development projects in those LDCs that were eligible for OPIC insurance. OPIC insures the petroleum investor against losses resulting from breach of specified contractual obligations by the host

government as well as outright expropriation. The insurance covers intangible exploration costs as well as equipment and structures. By the end of fiscal 1982 (ended September 1982) OPIC had issued 19 contracts for equity and loan insurance in the oil and gas sector covering more than $2.5 billion of investment in 11 developing countries.[3] During fiscal 1982, equity investment insurance was issued for oil and gas projects in seven countries—Pakistan, Tunisia, Oman, North Yemen, Turkey, Egypt, and Israel—plus a $50 million loan guarantee to a consortium headed by Phillips Petroleum for an offshore oil project in the Ivory Coast.

OPIC insurance is not available for investments in OPEC countries, so that it is not available in non-Middle Eastern countries such as Ecuador and Indonesia. In the past, OPIC insurance has not been available for investments in a number of important oil-importing LDCs, including Argentina, Brazil, and Chile, because of alleged violations of human rights in these countries. These restrictions were relaxed in 1982 and OPIC insurance is now available for about 100 LDCs, including those mentioned above. OPIC is not permitted to make available insurance on projects in countries with per capita incomes above $2,950 (in 1979 dollars).[4] Finally, some countries have been unwilling to sign agreements with OPIC that make them eligible for insured investments.

Conversations with officials of international petroleum companies suggest that the availability of OPIC insurance is unlikely to stimulate a substantial amount of exploration and development in the non-OPEC LDCs. Of the insurance policies issued by OPIC, only one has been issued to a major international oil company, namely, Gulf Oil Corporation, on its investment in Egypt. The political risk of expropriation or even contract violation on the part of the host government does not appear to be high during the exploration and early production period. In most cases demands for substantial contract

[3]Overseas Private Investment Corporation, *Annual Report 1982* (Washington, D.C., March 1982) p. 13.

[4]*Overseas Private Investment Corporation Amendments Act of 1981*, Public Law 97-65, 97th Cong. 1 sess. (October 16, 1981). Previously the per capita income limit was $1,000.

revision have occurred after the company has recovered its capital and is making satisfactory profits. OPIC insures only net unrecovered costs and not potential profits lost as a consequence of expropriation or forced contract revision. Compared with other risks and deterrents, many company officials believe the risks that can be insured by OPIC constitute a relatively minor constraint on petroleum investment in the oil-importing LDCs.

In the case of a high-risk investment in exploration, compensation by OPIC for the amount of the investment in the event of expropriation may not be sufficient to affect the decision to make the investment. For example, assume a petroleum company invests $50 million in an exploration program with a 1 to 5 probability of success. As was pointed out in chapter 4, the expected NPV that will justify the exploration investment must be five times the amount that would justify the investment if a commercial discovery were certain. In other words, the probability-adjusted NPV must be five times the NPV for successful discovery. Therefore, simply compensating the investor for his exploration outlay of $50 million would not be sufficient to induce the investment, since unless the exploration is successful, there would be no expropriation and no compensation of any kind. Moreover, the action of a host government is more likely to take the form of a contract violation that might reduce, say, from one-half to two-thirds the expected NPV for the project at the time of the investment. In such an event, the investor would have little incentive to terminate the investment and to claim compensation for the exploration outlay, even though OPIC's insurance policy provides for compensation in the event of a contract violation. Even with a contract violation, a successful discovery may be worth developing as contrasted with being compensated for the initial exploration outlay. An adequate compensation for expropriation or violation of contract terms might, therefore, be one that amounts to some multiple of the exploration outlays commensurate with the probability of commercial discovery. I doubt very much if it would be possible for OPIC to obtain statutory authority for insurance of this type and

the problems of determining the probability of success of an exploration venture would be difficult.

It is worth noting that the governments of other developed countries, especially Canada, Germany, and Japan, provide assistance to their petroleum companies in promoting overseas petroleum exploration and development. Several countries provide loan guarantees and interest subsidies on petroleum investments in foreign countries.

U.S. Foreign Tax Credits on Foreign Operations of U.S. Petroleum Companies

Since 1918, U.S. corporations have been able to credit foreign income tax payments against their U.S. corporate tax liabilities. Depending upon U.S. tax regulations, foreign tax payments may either be deducted as a cost in the calculation of U.S. taxable income or as a credit against U.S. tax liabilities. Assuming a 50 percent U.S. corporate tax rate, a $100 foreign tax payment reduces the U.S. tax liability by only $50 if treated as a cost, but by $100 if treated as a foreign tax credit. In granting petroleum concessions, mineral leases, or other types of contracts to U.S. firms, foreign governments have usually required royalties or other types of payments for the use of the resources; these payments differ conceptually from taxes on net income. Prior to 1950, most governments of developing countries did not impose income taxes, but relied mainly on royalties for their revenues from petroleum operations. In 1950, when Saudi Arabia sought more revenue from its oil properties, it decided, after consultations with American oil and tax experts, to levy an income tax on Aramco in lieu of increasing its royalties on oil production. Aramco requested a ruling from the IRS that the income taxes paid to the Saudi Arabian government be credited against the company's U.S. tax liability. There is substantial evidence that the U.S. Treasury opposed Aramco's request on the grounds that the income tax payment was in fact an increased royalty and that a U.S. tax credit simply increased Saudi revenues at the expense of the U.S. Treasury at no net cost to Aramco.[5]

After several years, the issue was resolved in favor of Aramco, largely as a consequence of intervention by the U.S. Department of State. The issue of what constituted a foreign income tax allowable as a foreign tax credit was further clouded during the late 1950s and early 1960s when Saudi Arabia and governments of other petroleum countries began basing taxable income on *posted* prices rather than on *actual* prices from the sale of oil and without regard to the actual net income of the petroleum companies. Although there were IRS rulings from time to time relating to the calculation of foreign tax credits, reasonably clear rulings or guidelines on what constituted creditable foreign tax payments were not made until after the passage of the Tax Reform Act of 1976.

U.S. Tax Rulings under the Tax Reform Act of 1976

In a *News Release* issued July 14, 1976, the IRS set forth certain general principles with respect to the creditability of foreign taxes paid by taxpayers engaged in the extraction of mineral resources owned by foreign governments. These principles were as follows:

(1) The amount of income tax is calculated separately and independently of the amount of the royalty and of any other tax or charge imposed by the foreign government.

(2) Under the foreign law and in its actual administration, the income tax is imposed on the receipt of income by the taxpayer and such income is determined on the basis of arm's length amounts. Further, these receipts are actually realized in a manner consistent with U.S. income taxation principles.

(3) The taxpayer's income tax liability cannot be discharged from property owned by the foreign government.

(4) The foreign income tax liability, if any, is com-

puted on the basis of the taxpayer's entire extractive operations within the foreign country.

(5) While the foreign tax base need not be identical or nearly identical to the U.S. tax base, the taxpayer, in computing the income subject to the foreign income tax, is allowed to deduct, without limitation, the significant expenses paid or incurred by the taxpayer. Reasonable limitations on the recovery of capital expenditures are acceptable.[6]

Indonesian Production-Sharing Contracts

The above ruling had special significance for the Indonesian production-sharing contracts which had been negotiated between Pertamina and a number of U.S. petroleum companies since the 1960s. Prior to 1976, U.S. petroleum companies had been able to obtain a U.S. foreign tax credit equal to that portion of Pertamina's share of the output used to discharge the U.S. contractor's income tax obligation to the Indonesian government. However, payments by U.S. companies to Pertamina under production-sharing contracts violated the principles of the IRS as stated in its announcement of July 14, 1976. Under the production-sharing contracts, income tax payments to the Indonesian government were made by Pertamina and not directly by the foreign investor. Furthermore, the IRS argued that there was not a clear separation between the amount of income tax and the amount of royalty imposed by Indonesia. Therefore, in a U.S. Treasury Department *News Release* of April 8, 1976 (prior to the statement of principles in its *News Release* of July 14, 1976), the IRS took the position that "the share of production received by the foreign government is in substance a royalty in entirety and not eligible for the foreign tax credits under sections 901 and 903 of the Internal Revenue Code."[7] Thereafter, there was a dispute within the U.S. government regarding whether U.S. companies operating under production-sharing contracts in Indonesia should be subject to retroactive taxation, or, indeed, whether they should be forced to renegotiate existing contracts with Pertamina as a

---

[5]See *Foreign Tax Credits Claimed by U.S. Petroleum Companies* (Hearings before Subcommittee of the Committee on Government Operations, House of Representatives, 95th Cong. 1 sess. (September 26, 1977) (Washington, D.C., Government Printing Office, 1977) p. 2.

[6]Ibid. p. 345.
[7]Internal Revenue Service, *News Release* (Washington, D.C., U.S. Department of the Treasury, April 8, 1976).

condition for receiving any U.S. income tax credits. (The U.S. Department of State took the position that the proposed IRS denial of all tax credits arising from the Indonesian production-sharing agreements would tend to discourage U.S. petroleum investment in Indonesia.) The issue was finally resolved by the renegotiation of all the Indonesian production-sharing contracts in line with IRS rulings. As noted in the discussion of the Indonesian production-sharing contracts (chapter 5), all contracts negotiated after 1976 provided for direct tax payments to the Indonesian government and were based on the net income of the contractors, rather than having the taxes paid by Pertamina.

The Peruvian Contracts

A somewhat similar foreign tax problem arose in the case of the contracts negotiated by the Peruvian state agency, Petroperu, with U.S. petroleum firms, only two of which (Belco Petroleum and Occidental Petroleum) found any oil and made any profits. Under the contracts negotiated between 1971 and 1978 (see chapter 6) the contractor split the petroleum output (usually on a 50-50 basis) with Petroperu, and Petroperu paid the Peruvian income tax due on the contractor's operations. However, unlike the Indonesian production-sharing contracts, the contractor bore all the development and operating costs and there was no recovery of these costs before the production split.

Since it was clear that the Peruvian contracts also violated the new IRS regulations with respect to the crediting of foreign tax payments, the two American companies, led by Belco, reached an informal agreement with officials of the Peruvian government for a revision of the existing contracts, according to which the companies would pay a 40 percent tax on *gross* income directly to the Peruvian government and a royalty to Petroperu. According to the proposed agreement, legally the companies would have the option of operating under the proposed arrangement or of being subject to the regular Peruvian tax of 68.5 percent on net income applicable to foreign mining and petroleum companies. The Peruvian government officials favored the proposed arrangement of a royalty plus a regular tax on net income since it simplified the problem of determining net income after operating costs and various capital allowances. Belco Petroleum officials argued that the 40 percent gross tax would provide less tax credits against U.S. taxable income than if the regular Peruvian net income tax had been applied. In a brief dated August 31, 1978, Belco Petroleum's tax lawyer argued that Section 903 of the IRS code permits crediting a gross income tax "in lieu of" a tax on net income provided certain criteria were met, and that the proposed Peruvian tax law together with other Peruvian tax legislation fully met these criteria.

On August 3, 1979 the IRS issued a ruling in the form of a letter to Belco Petroleum stating that the proposed gross income tax would be regarded as a tax in lieu of an income tax within the meaning of section 903 of the IRS Code. The regulations on which the IRS ruling was based were stated in an *IRS News Release* entitled "IRS Issues Regulations on Foreign Tax Credits" (U.S. Department of the Treasury, November 12, 1980). These regulations established two circumstances under which a foreign tax other than a tax on net income may be credited against U.S. tax liabilities: "(1) The foreign country imposes an income tax which does not treat extractors significantly differently from other taxpayers; or (2) extractors are subject to a tax that is not the foreign country's general income tax but the amount paid is comparable to what would have been paid under the country's general income tax."

Unfortunately, the proposed Peruvian gross tax never went into effect and the new Peruvian contract legislation, which became effective in 1980, provided for a tax on net income. The delay of nearly a year by the IRS in issuing the ruling on the proposed Peruvian gross tax, despite repeated requests by officials of Belco, the DOE, and the Peruvian government, delayed negotiations for new contract areas with U.S. petroleum companies. There is evidence that had the new IRS regulation been issued more promptly, the petroleum companies could have negotiated new contracts which would have been more favorable to them than the ones subsequently negotiated in April 1980 under the new Peruvian Decree Laws.[8]

The Guatemalan Contracts

Contracts negotiated by the Guatemalan government with U.S. petroleum companies under a model contract for petroleum operations issued in January 1978 are also probably at variance with IRS regulations for crediting foreign income tax payments, although to my knowledge the IRS has not issued a ruling on the creditability of these taxes. The Guatemalan contracts are essentially production-sharing contracts with no provision for cost recovery. Contractors pay a graduated net income tax directly to the Guatemalan government, but the government then reimburses the contractor for the total amount of income taxes paid (see chapter 7). This arrangement appears to violate the principle of complete separation between royalties and income and withholding taxes on profits remitted abroad. However, as of late 1982, company officials with petroleum investments in Guatemala stated that the question of creditability of Guatemalan taxes had not been resolved with the IRS.

The Egyptian Contracts

Several U.S. companies are operating in Egypt under PSCs that provide for income tax liabilities of their Egyptian subsidiaries to the Egyptian government to be paid by Egypt's government-owned enterprise, Egpc. In August 1980 the IRS issued a favorable ruling with respect to a proposed restructuring of the Egyptian tax regulations and in August 1981 a U.S. company received a "private letter" from the IRS permitting that company's tax liabilities to the Egyptian government to be credited against its U.S. tax liabilities, despite the fact that the tax liability to the Egyptian government was paid by Egpc.[9] However, the creditability of the company's 1981 and subsequent years' taxes

was made subject to certain changes in the Egyptian tax regulations, which I understand have been made. The IRS ruling set forth in the "private letter" in effect regards the Egyptian tax arrangement as consistent with section 903 of the IRS code, which permits crediting a gross income tax "in lieu of" a tax on net income. However, unlike the ruling in the Belco Petroleum case mentioned above, the Egyptian tax is not a gross tax paid by the U.S. company, but a net income tax paid by Egpc. Other aspects of the Egyptian tax arrangement are apparently in accord with the principles set forth in the IRS statement of July 14, 1976, summarized above. All of this appears to suggest that there is some flexibility in the application of the principles set forth by the IRS in its July 14, 1976 statement, and that the matter of creditability under PSCs that provide for income taxes to be paid by an agency of the host government is being dealt with on a case-by-case basis.

Tax Equity and Fiscal Responsibility Act of 1982 (TEFRA)

The Tax Equity and Fiscal Responsibility Act of 1982 made some additional changes relating to the ability of U.S. petroleum companies to credit foreign taxes against their U.S. tax liabilities. Perhaps the most important change was the repeal of the "single country loss rule" provided in earlier legislation, which states that a net extraction loss in any single country need not be taken into account when computing the tax credit limit for total foreign extraction income.[10] A loss in a single country therefore would not reduce the available overall foreign extraction tax credits from operations in other countries. The repeal of the per country

---

[8]This statement is based on the author's knowledge of and involvement in the discussions.

[9]The author was given a copy of this "private letter" by the U.S. company.

[10]Currently U.S. tax law requires that petroleum companies aggregate their taxable foreign income from all countries for purposes of determining allowable foreign tax credits. Previously they were permitted to calculate their tax credits on the basis of individual country operations. For an analysis of the impact of the "per country limitation" versus the "overall limitation," see Daniel D. Frisch, *Issues in the Taxation of Foreign Source Income*, Working Paper 798 (Cambridge, Mass., National Bureau of Economic Research, November 1981).

extraction loss rule requires that a net extraction loss in one country be deducted from the extraction income in other countries when computing the amount of creditable extraction taxes. Although the effect of this change in the law, in combination with other changes made by the 1982 law, on the ability of U.S. petroleum companies to offset foreign taxes against U.S. tax liability is somewhat ambiguous, the National Petroleum Council has argued that the change reduces the incentive of U.S. petroleum companies to invest in exploration abroad.[11]

Do U.S. Tax Laws Favor Investment Abroad?

U.S. income tax laws favor investment abroad over investment in the U.S. in two respects. First, profits earned abroad through a subsidiary corporation are not subject to U.S. taxation until remitted to the U.S. parent corporation.[12] Thus, undistributed profits of subsidiaries whose foreign tax liability is below the U.S. tax enjoy a tax saving to the extent of that differential over U.S. companies operating in the United States (or operating abroad through branches). Second, in calculating the amount of foreign tax that may be claimed as a credit, foreign income taxes paid on petroleum operations in all countries are aggregated. If a company has an operation in a country where the income tax exceeds the U.S. tax, it has an incentive to invest in a foreign country where it will incur an income tax below the U.S. tax in order to use its "excess" credits. Investment in the U.S. will not serve this purpose because the credit is available only against U.S. taxes on foreign earned income. For example, if an oil company has 100 income in country A taxed at 60 percent, it may credit 46 of foreign tax (46 percent of 100), leaving it 14 of excess credit. If the company earned income in country B with a 32 percent

tax, the company would have 92 U.S. tax credits (46 percent × 200) against the total foreign tax of 92. The company is thus encouraged to use its excess credits by investing in low-tax foreign operations. Under present law, the excess credits can be used only against income from oil-related activities. To some extent this effect can be viewed as an incentive to oil companies to explore new areas, such as non-OPEC LDCs. However, in most non-OPEC LDCs, the corporate income tax is as high or higher than the U.S. corporate tax.

Prior to January 1983, when the TEFRA amendment went into effect, an excess tax credit could be used to offset foreign losses such as those generated by oil drilling expenses. For example, if an oil company had 100 income in country A and 40 of losses in country B, it could credit 46 of foreign tax (46 percent of 100), but its U.S. tax would be only 27.6 (46 of 60), leaving it 12.4 of excess credit. However, under the new regulations, the U.S. tax would be 46. Under the previous tax credit arrangement, petroleum companies might be induced to undertake exploration activities abroad since the losses could be used to reduce aggregate foreign income taxable in the United States.

Tax Treatment by Other Developed Countries

Other developed countries have been more liberal than the U.S. in their tax treatment of foreign income of their petroleum companies. For example, France, Germany, and Japan, either specifically or through a variety of special regulations and administrative practices, effectively exempt foreign-source petroleum income from any domestic taxes. Germany and Japan allow full deduction against domestic-source taxable income for overseas exploration outlays, and, in addition, provide loans for exploration which are forgiven if the venture is unsuccessful.[13] Nevertheless, U.S. petroleum firms continue to be the most active in petroleum exploration in non-OPEC LDCs. U.S. firms drilled over 26

[11] See The National Petroleum Council, *Third World Petroleum Development: A Statement of Principles* (Washington, D.C., 1982) pp. 18–19. The Council is a Federal Advisory Committee to the U.S. Secretary of Energy.

[12] There are certain exceptions to this principle contained in the Internal Revenue Code's anti-tax-haven rules.

[13] Comptroller General, *Potential for Diversifying*, p. 67; see also National Petroleum Council, *Third World Petroleum*, pp. 19–20.

percent of all the wells drilled in developing countries in 1980, including those drilled by GOEs in these countries.[14]

Efforts to Achieve Tax Neutrality

IRS tax rulings as applied to petroleum companies in recent years have undoubtedly eliminated certain abuses which were costly to the U.S. government and favored foreign operations by U.S. petroleum companies. For example, some of the OPEC countries based their income taxes on artificial posted prices, in contrast to actual prices received by the petroleum producers, which meant that taxable income was higher than actual income. Also, the insistence of the IRS that a reasonable proportion of the taxes paid by the petroleum companies constitute royalties, rather than using the income tax to include both income taxes and royalties, is in line with the concept of neutrality of treatment of income taxes at home and abroad. The IRS also ruled that a U.S. corporation operating abroad is permitted to credit income taxes paid to foreign governments only against U.S. income tax liability on that same income; the credits may not be used to offset tax liabilities on the corporation's domestic income. IRS rulings are also designed to ensure that tax credits offset only the U.S. tax on foreign income by setting an upper limit on the amount of credit claimed. Generally this limitation prevents the crediting of a foreign tax which is in excess of the U.S. tax due on the same income (currently 46 percent).[15]

Alternative Proposals for Taxing
Foreign Income

In general, the current approach of the IRS has been to avoid encouraging or discouraging petroleum production abroad through the U.S. tax system. However, other options have been put forward by members of Congress and other groups. One is the complete elimination of foreign tax credits, and another, which would also discourage foreign operations, is to impose U.S. income taxes on foreign subsidiary income, whether or not the income is remitted to the U.S. parent, i.e., eliminate tax deferral. The argument advanced for such changes is that they would stimulate domestic exploration activity by making investment in marginal foreign fields less attractive.

Another approach would be to liberalize the tax treatment of foreign income from petroleum operations in certain areas where the U.S. government would like to encourage petroleum exploration and production. Presumably this would have to be done by means of bilateral tax agreements with the countries in which the U.S. government desired to encourage U.S. petroleum activity.

One tax arrangement favoring U.S. petroleum investment in a particular country would be to permit U.S. companies to elect the "per country" tax credit limitation on taxes paid in that country. Another would be to provide a more generous interpretation of eligibility of foreign taxes for U.S. tax credits in the case of countries such as Guatemala that have been unwilling to adjust their own production-sharing tax or royalty system in line with IRS regulations. This could be done without a change in the amount of foreign taxes otherwise creditable against U.S. tax obligations. Finally, the U.S. government could provide investment tax credits for petroleum investment in non-OPEC LDCs. This action is favored on a selective basis by the National Petroleum Council.[16]

The value of liberalizing the treatment of foreign tax credits for petroleum companies is debatable; the analysis is complex and the outcome is uncertain. Liberalizing foreign tax credits could benefit the U.S. oil industry, the host country, the consumer, or some combination of all three. The same is true regarding the effects of abolishing foreign tax credits entirely. In the case of the major Middle Eastern producing countries

[14]Ibid., appendix D.
[15]For a discussion of current IRS tax credit policy, see Report by the Comptroller General, *The Foreign Tax Credit and U.S. Energy Policy* (Washington, D.C., U.S. General Accounting Office, EMD/80/86, September 10, 1980).

[16]National Petroleum Council, *Third World Petroleum*, p. 21.

that have either nationalized their petroleum industry or in various ways limited exploration and production by foreign firms, foreign tax credits are unlikely to have any impact on U.S. petroleum investment in these countries. The same is true of Venezuela, which has nationalized its petroleum industry. Even in the case of countries such as Indonesia, where there is active competitive bidding for contracts, there is a real question whether petroleum exploration and development would be greatly affected by either liberalization or denial of foreign tax credits. In the latter case, the companies might simply bid somewhat less for the available contracts. Since most countries seek to obtain the maximum rents consistent with the level of foreign investment they want to attract, it could be argued that either the elimination of foreign tax credits or their liberalization would be absorbed by the host countries. On the other hand, it seems likely that liberal foreign tax credits, including the right to deduct all losses on a per country basis against U.S. tax liabilities on domestic income, would encourage exploration by U.S. companies in high-risk areas such as Brazil, Guatemala, Argentina, Chile, Bolivia, or Peru. Nevertheless, in most of these countries a favorable change in their domestic tax legislation would undoubtedly provide a much greater incentive to U.S. petroleum companies than a liberalization of U.S. foreign tax credits. However, there are often political limits to the degree to which taxation of petroleum companies can be liberalized in many third world countries.

Conclusions

In conclusion, the elimination of foreign tax credits or tax deferral would be unwise since it might have a deterrent effect on U.S. petroleum investments, especially in non-OPEC countries. On the other hand, any liberalization of foreign tax credits or an expansion of the investment credit should be applied on a selected basis to countries in which the U.S. has a special interest in promoting petroleum investment. However, tax subsidies should not be granted unless the social benefits to the U.S. clearly exceed the social costs, and it is not clear that subsidizing any foreign petroleum production will provide sufficient social benefits to justify the costs. The Department of the Treasury's "neutral" approach to U.S. investment at home or abroad may be the optimum policy.

12

Promotion of Petroleum Development in the Oil-Importing Developing Countries by International Agencies

International agency assistance for petroleum exploration and development in the LDCs has been made available mainly from the United Nations Development Program (UNDP) and from the World Bank Group, with a few loans from the Inter-American Development Bank (IADB). Modest grants for mineral surveys in developing countries have been made by the UNDP for more than two decades. The UNDP has supported geological surveys and has provided training for employees of GOEs and assistance on petroleum legislation and in negotiations with petroleum companies. However, the UNDP has provided little financial assistance for exploration because such activities are too costly in view of that agency's limited budget.[1]

Interest in substantial support for petroleum exploration and development in the OIDCs dates from the early 1970s, and was stimulated by the

sharp rise in petroleum prices that has placed a severe burden on the balance of payments of these countries. A number of proposals have been put forward in the UN for establishing facilities for making large grants or loans to the OIDCs for exploration and development of both fuel and nonfuel mineral resources. These proposals generally emphasize development by state petroleum and nonfuel mineral enterprises and reflect in part the principle of "full permanent sovereignty of every state over its natural resources, including the right of nationalization or transfer of ownership to its nationals," which has been an important element in the New International Economic Order (NIEO). The UN General Assembly resolution of May 1, 1974 provided that "competent agencies of the UN meet requests for assistance from developing countries in connection with the operation of nationalized means of production."[2]

The U.S. government, as well as those of other industrial nations, has reacted negatively to proposals for the creation of a special UN

---

[1]See UN Development Program, *Energy Fund for Exploration and Preinvestment Surveys* (New York, United Nations, UNDP/438, April 2, 1980). The UN Revolving Fund for Natural Resources Exploration was originally designed to finance petroleum exploration as well as exploration in the mining sectors, but the petroleum exploration activity was abandoned because of limited resources.

[2]*General Assembly Resolution 202 (S/VI)* (New York, United Nations, May 1, 1974).

agency for financing the exploration and development of energy resources in the OIDCs, mainly on the ground that such financing should come from private sources. However, the U.S. government has supported World Bank financing of petroleum activities provided such financing complemented rather than substituted for private petroleum exploration and development. In July 1977 the executive directors of the World Bank approved a program calling for expanded lending for the development of fuel and nonfuel mineral resources of member countries.[3]

The World Bank Group Program

In a report of August 1980, the World Bank outlined an assistance program for energy development for fiscal years 1981–85 that included (a) technical assistance for energy sector review; (b) promoting oil and gas exploration; and (c) investment in oil and gas production. The report states that in exploration, "we shall seek to maximize participation by private companies which have traditionally provided risk capital and necessary knowhow. However, in some cases . . . international oil companies may seek the presence of the Bank at the exploration stage or the Bank may be requested to participate in financing exploration programs undertaken jointly by private and national oil companies, or by national oil companies alone."[4] In financing oil and gas production, the 1980 World Bank report emphasized financial and technical assistance to national oil companies, since presumably international petroleum companies are generally capable of financing the development of oil fields in which they are operating under concessions or other forms of contracts with national governments.

The lending program envisaged by the World Bank for FY 1981–85 totals about $4.0 bil-

lion—$1.0 billion for oil and gas predevelopment; $1.8 billion for oil development; and $1.2 billion for gas development. The total project costs are estimated at $11.8 billion, so that the Bank would be expected to finance about one-third of the total project costs. However, the Bank has tentatively identified an additional $4 billion in lending for this period, but stated that the additional funds would need to come from special sources contributed to a new bank affiliate or facility organized for this purpose. Therefore, in late 1980 the president of the World Bank, Robert McNamara, proposed a new energy facility with capital of about $10 billion to be subscribed by the governments of the industrialized and OPEC countries. Ten percent of the capital would be paid in and the remainder would be borrowed on the security of the unpaid subscribed capital. However, early in 1981 the Reagan administration informed the World Bank's executive board that it would neither support the creation of, nor participate in, the proposed energy affiliate.[5]

World Bank Group Financing During 1977–82

Following the World Bank's executive board decision to finance petroleum projects in July 1977, the World Bank and the International Development Association (IDA) made nearly $1.7 billion in loans through June 30, 1982 for oil and gas exploration and development and related infrastructure to four non-OPEC oil-exporting countries and more than 25 OIDCs. These loans fall into several categories. The largest portion, nearly $1 billion, was made to national governments or GOEs for exploration and development projects in which there was no domestic private or foreign equity investment. Another portion totaling about $110 million was made to GOEs to enable them to participate in joint ventures with foreign companies. About $350 million was made to governments or GOEs for infrastructure (mainly gas pipelines); over

[3]See World Bank, *Annual Report 1978* (Washington, D.C., 1978) pp. 20–22, for a summary of the program and its rationale. Prior to that time, the World Bank had not financed petroleum production projects.
[4]World Bank, *Energy in Developing Countries*, pp. 74–75.
[5]"Reagan Asks World Bank to Scrap Plans for Double Lending for Energy Projects," *Wall Street Journal*, August 4, 1981, p. 10.

$100 million was made for financing private domestic oil operations; and about $100 million went for "exploration promotion." Of the loans made to governments or GOEs for projects without private equity participation, over half ($550 million) consisted of loans to India. In addition, there were loans in this category to Turkey, Peru, Morocco, Argentina, Egypt, and Tanzania.[6]

Most of the exploration promotion loans made by the World Bank and IDA have been made to non-oil producers, and by June 1982 such loans had been made to 16 countries, most of which were for less than $5 million. The principal purpose of these loans is to accelerate the offering of contracts to international petroleum companies. The projects financed include geological surveys or reconnaissance seismic drilling; technical assistance in training nationals; advice on the negotiation of contracts with international petroleum companies; and the reduction of political risk through a variety of arrangements ensuring a World Bank "presence" during the exploration and/or development phases of foreign contractor activity.

An example of a World Bank loan to support joint participation by state enterprises with foreign oil companies for exploration and development is the $102 million loan to Ivory Coast to enable Petroci, the national oil company, to participate in the exploration and development of offshore petroleum with a consortium of oil companies headed by Phillips Petroleum.[7] The World Bank also made a commitment to consider financing the Pakistan government's share of a joint exploration-development program with Gulf Oil, in which Gulf would serve as operator and pay 85 percent of the costs of exploration. Should commercial discoveries result, Gulf and Pakistan Oil and Gas Development Corporation would develop the field on a 50-50 basis.[8]

The International Finance Corporation (IFC), an affiliate of the World Bank that makes loans and investments to projects involving private equity, has made a few loans in support of domestic private investors in petroleum operations or joint ventures involving state enterprises and private investors. For example, in March 1981 the IFC agreed to participate with a group of external banks in the financing of an exploration and development project in Colombia involving a private group, Petroleos Colombianos (Petrocol). The IFC would have an equity interest of $3.4 million in Petrocol, as well as participating to the extent of $3.2 million in a $12 million loan to that company. The IFC also made a loan of $4.1 million for petroleum development to a joint venture involving the government of Zaire and a consortium of private oil companies, and in 1982 the IFC made a $300,000 equity investment in a joint venture for petroleum development involving the government of Sudan and Chevron Oil, which has made oil discoveries in Sudan.[9]

Inter-American Development Bank Assistance

The Inter-American Development Bank (IADB), in a major policy innovation, began making loans for oil and gas exploration in 1980. Its first such loan was to Jamaica ($23.5 million) to help finance onshore oil and gas exploration.[10] Another loan for $16 million was extended to help Bolivia drill fourteen test wells and five gas fields.[11] In 1981 the IADB approved a $5.3 million loan to help Peru finance marine seismic studies for petroleum exploration in an offshore area and a $35 million loan to Brazil for a similar purpose.[12] During 1981 the IADB also made three loans for marine seismic surveys to Brazil, Colombia, and Peru totaling about $50 million. In 1982 the IADB

[6]For sources of these data, see World Bank, *Annual Reports* 1978 through 1982 (Washington, D.C.) and Treasury, *World Bank Energy Lending*, pp. 31–32.

[7]World Bank, *Annual Report 1982* (Washington, D.C., 1982), p. 107.

[8]Ibid. p. 38.

[9]International Finance Corporation, *Annual Report*, 1979, 1981 and 1982 (Washington, D.C.).

[10]Inter-American Development Bank, *Annual Report 1980* (Washington, D.C., 1980) p. 32.

[11]"Developing Nations Receive Help to Increase Fuel Supplies," *New York Times*, October 2, 1980, p. 27.

[12]Inter-American Development Bank, *News Release* (Washington, D.C., April 1, 1981).

made three loans totaling $134 million to Bolivia's state petroleum company for oil exploration and development.[13] All of the IADB petroleum loans have been made to GOEs, but in some cases they have served to promote the negotiation of contracts with private companies.

Evaluation of International Agency Assistance

A major question arises as to whether the activities of the World Bank group and the IADB in financing petroleum exploration and development will on balance serve as a catalyst for mobilizing private petroleum investment in the OIDCs, or whether their activities will simply reduce the incentive of governments to attract private investment. The case in which the World Bank has committed itself to financing a joint venture involving Gulf Oil and the Pakistan state petroleum enterprise qualifies as a clear incentive to private investment. Gulf Oil officials have indicated to me that this arrangement provides a measure of security for their investment. The loans are clearly designed to promote exploration by international petroleum companies. The IFC investments in joint ventures involving private capital also serve this purpose.

A U.S. Department of the Treasury staff report argues that, contrary to the stated purpose of the World Bank's petroleum loans to "catalyze" private investment flow, the bulk of the Bank's loans have been made to national oil companies with no inducement to private capital. In fact, the Treasury staff report argues that such loans are "likely to have displaced private capital investment."[14] It further states that the terms on which governments have obtained World Bank funds were superior to those available to private companies, whether in the form of debt capital or equity. The Treasury report's criticism is mainly on grounds that, if the World Bank's loans were employed to induce private invest-

ment, "total oil and gas exploration and development investment would increase significantly."[15] No objection was made to the exploration promotion loans, since they are designed to induce exploration by foreign petroleum companies and to facilitate the negotiation of contracts.

International petroleum company officials argue that, given favorable contracts, they are willing to undertake exploration and provide the financing for development almost anywhere in the world. Many also express the concern that international agency financing of state petroleum enterprises will retard petroleum exploration and development in the OIDCs since it provides an alternative to negotiating contracts with international petroleum companies that have both the technical ability and the financing for the efficient development of petroleum resources.[16]

On the other hand, for some governments there are political constraints on negotiating contracts with private petroleum companies so that in the absence of international agency assistance, adequate petroleum investment will not take place. Moreover, it is argued that international petroleum companies are not interested in investing in countries with small domestic markets and poor prospects for finding sufficient reserves to make a petroleum exporting industry profitable. However, this argument is denied by petroleum company officials who point to a number of cases of petroleum exploration and development investments by international oil companies in countries that are unlikely to become petroleum exporters. The question may be raised whether the development of a national petroleum industry that would be unprofitable for an international company would be economical (in social benefit-cost terms) for a state petroleum enterprise run with the aid of foreign financing and technology. There may be some circumstances in which this is true, especially if some of the risk is borne by international

[13]Inter-American Development Bank, *Annual Report 1982* (Washington, D.C., 1982), pp. 47–48.
[14]Treasury, *World Bank Energy Lending*, pp. 33–34.
[15]Ibid. p. 63.
[16]This is, in general, the position taken by the National Petroleum Council in *Third World Petroleum Development*, pp. 24–25.

assistance agencies.[17] There seems little doubt that GOEs are increasing their technical capacity and that they are able to employ effectively both technical and financial assistance available from the international agencies for developing their petroleum resources. For example, Petrobras has been successful in expanding Brazil's output and reserves in recent years[18] and India's GOE also has been quite successful in both onshore and offshore operations.

There is little criticism of World Bank loans and technical assistance for improving the technical capacities of GOEs to conduct surveys and preliminary exploration and to formulate and carry out a national petroleum production program, including negotiating with international petroleum companies. Loans for this purpose constitute an effective use of international agency capital. Some of the projects financed have yielded benefits in the form of geological information and actual oil discoveries, and in some cases have led to the negotiation of contracts with international oil companies.[19]

The basic issue arises over the appropriateness of large loans to GOEs for intensive exploration and development which international oil companies would be willing to undertake if suitable contract terms were offered. This question should not be decided on ideological grounds, i.e., private versus public ownership and production. Some of the international oil companies, such as Elf Aquitaine, are themselves government enterprises. Nor should the question be decided on the basis of the argument that public international financing of GOEs constitutes unfair competition with U.S. or other foreign private enterprises. Rather, it should be decided on the basis of the most effective use of the limited amount of international public assistance capital. If international companies are willing to provide billions of dollars in risk capital for exploration and development of oil re-

sources, should the limited World Bank and IDA resources be used for this purpose? There is also the question of efficiency involved since some GOEs have a poor record in terms of efficient operation and management, or lack the technical capacity for certain operations, such as deep sea offshore exploration and production. However, this does not apply to all GOEs. International assistance agencies should concentrate their loans on projects designed to mobilize foreign private sources of capital and technology. The principle suggested does not mean the GOEs could not play an important role in petroleum exploration and development, even if the bulk of the capital for a project came from international petroleum companies. Modern contracts usually provide for a substantial amount of control and surveillance by governments, regardless of the contractual form, and joint ventures and other types of arrangements involving joint policy determination and management participation by private companies and GOEs are quite common.

International assistance agencies can play an important role in reducing the risk exposure of international petroleum companies through their association with projects as lenders or, in the case of the IFC, as a minority equity investor. The presence of these institutions can help to assure the maintenance of contract obligations. In this way the international institutions can perform a catalytic function by mobilizing international capital and technology, while limiting their own financial participation.

It may be noted that the National Petroleum Council report cited above is critical of a direct involvement of the World Bank in contract negotiations on the grounds that "in some situations the Bank's impartiality may be jeopardized, and it could be forced into taking partisan positions in normal commercial relationships in negotiations that might better have been conducted on a bilateral basis between the private petroleum company and the host country. OIDCs seeking to maximize their exploration potential will often find they are best served in such efforts by working within the framework of competitive negotiations involving several private petroleum companies

[17]This also raises the question of the most effective use of foreign aid in terms of promoting the welfare of developing countries.

[18]Petrobras operates in other countries as an international oil enterprise as well as in Brazil.

[19]See World Bank, *Annual Report 1982*, pp. 39–40.

and through the use of independent consultants as needed.''[20]

I am not in sympathy with the NPC report on this issue. First, the private petroleum companies are well able to take care of themselves and World Bank involvement in contract negotiations does not imply that the World Bank staff

[20]National Petroleum Council, *Third World Petroleum Development*, p. 24.

will be biased against the legitimate interests of the private companies. Unless the private companies are attracted to the project or unless they succeed in realizing their objectives, the entire project is unlikely to be successful. The fact is that international agency participation can provide a substantial measure of security for the private petroleum companies by greatly reducing the likelihood of contract violations.

13

Conclusions

Petroleum Potential of the Non-OPEC LDCs and the Adequacy of Exploration Activity

Although petroleum geologists differ in their estimates of potential petroleum resources in the non-OPEC LDCs, and particularly in the OIDCs, estimates made by the World Bank staff and by the U.S. Department of Energy indicate ultimately recoverable oil resources of five to ten times the present estimated proved oil reserves of the OIDCs. Although these potential oil reserves are heavily concentrated in the oil-producing OIDCs, there are a number of other OIDCs with favorable oil prospects. There is substantial evidence that a considerably higher level of exploration activity is warranted in most, if not all, OIDCs.

An *adequate* amount of exploration is not measured by the number of wildcat wells per 10,000 square miles based on exploration efforts in large oil-producing countries such as the United States. Adequacy is determined by the exploration that is economically warranted from

the geologic knowledge gained at each step in the process, ranging from geological and geophysical surveys to the drilling of wildcat wells. The decision to proceed with additional exploration outlays is a function of accumulated knowledge, costs, and estimated probabilities of success, and such decisions can be made only by technically competent and experienced petroleum firms, whether publicly or privately owned. In some countries GOEs may be constrained by the lack of technically trained personnel or by funds for pursuing an adequate level of exploration. International petroleum companies may be attracted to exploration in a particular country only if a certain amount of geologic knowledge has been accumulated and the results of preliminary investigations are promising. In addition, of course, they will require favorable contract conditions and a favorable political and economic environment. Inadequate exploration in some countries is a consequence of the failure of government to organize and fund a geological investigation program and a group of trained officials capable of

134

conducting negotiations with prospective petroleum companies within the petroleum laws of the country.

General Observations on Modern Petroleum Agreements

The modern mineral agreement negotiated with host governments (whether concerned with petroleum or other minerals) is an anomaly. It is nominally a contract to conduct certain activities under host government supervision, but since the risk and financing are borne by the contractor, it is a foreign equity investment. The foreign investment is basically the contract itself, since in most agreements title to the resources in the ground and even to the equipment and installations is held by the host government or reverts to the government at the termination of the contract. Moreover, the contracts have not been regarded by most host governments as covenants to be honored throughout their term, but rather as a framework for more or less continuous negotiation as the relative bargaining power and opportunities created by new conditions change in favor of the government.

In the major OPEC countries, such as Saudi Arabia and Kuwait, the position of the original contractor has diminished to that of a technician and buyer of the oil produced, with little or no control over the amount of oil produced or received, or over the price that must be paid to the host government for the oil. The position of petroleum companies operating in non-OPEC countries, as well as in some of the smaller OPEC producers such as Ecuador and Indonesia, is much stronger since most of these countries need substantial exploration and development to expand their reserves and output, and the governments lack the financial and technical resources to undertake investments on an adequate scale. The situation in these countries is far different from that of Saudi Arabia, for example, which in order to maintain the world price of oil is producing at a level well below that warranted by proved reserves, and large additional

Saudi reserves could probably be established without high-risk exploration. Nevertheless, the modern petroleum contract, whether it takes the form of a PSC, a joint venture, a risk-service contract, or a concession contract, prescribes in considerable detail the activities of the contractor, including the number of exploration wells he must drill or exploration outlays he must make, the development and production of any discoveries, and frequently the price and other conditions for the sale of the oil.

Since in recent years world oil prices have been quite high relative to operating and capital costs actually incurred for development and production of successful oil discoveries,[1] the governments of host countries have claimed the bulk of the gross revenues produced, and the share of net revenues claimed by governments ranges from 85 to 95 percent. A profit share of 10 to 15 percent on the output of a large field may yield a very attractive IRR to the petroleum company if no account is taken of risk. But the economics of investment decision making in a high-risk environment requires an expected or probability-adjusted IRR that takes into account several categories of risk. These include exploration risk with low probabilities of finding large fields and higher probabilities for smaller fields; risks at the development stage (since not every production well drilled in a known field will yield oil economical to produce); and a variety of other risks relating to costs and world oil prices, plus the political risk of contract violations. Risk perceptions (as well as the degree of risk aversion) differ among petroleum companies and the companies and host governments have differing evaluations regarding risk and compensation for risk. Differing contract terms designed to achieve the same ratio of net returns to the contractor and the host government will yield substantially different expected IRRs for the same petroleum field, and actual IRRs will differ greatly with the size of the field and other conditions for the same set of contract terms.

---

[1]The full cost of producing crude oil in the OIDCs in 1980 was estimated to be in the range of $6 to $15 per barrel (in 1980 dollars). World Bank, *Energy in Developing Countries*, p. 9.

Effects of Contract Terms on the Level of Petroleum Exploration

Many OIDCs with petroleum potential did not seek to attract foreign petroleum companies until after the sharp rise in oil prices in 1974, and their initial efforts have frequently brought relatively little exploration, in considerable part because they have not offered sufficiently attractive terms to private investors. Countries with small proved reserves have often patterned their agreements along the lines of those negotiated by countries with substantial reserves and favorable geologic conditions for large discoveries. High-risk exploration with low probability of success requires generous terms in order to provide substantial rewards for successful ventures. Moreover, since newly discovered fields in regions not producing oil are likely to be relatively small, while exploration and development costs are quite high, the expected present value of petroleum projects before taxes may not be exceptionally high even if contract terms are quite generous. Once important discoveries are made, new contracts more favorable to the government can be negotiated on tracts in the same general area.

Governments have generally done a poor job in structuring their tax and revenue or production-sharing arrangements for achieving maximum exploration activity and maximum efficiency in the development of discoveries. They have often patterned their petroleum laws and model contracts on those of countries with different petroleum characteristics, or have been more concerned with maximizing *potential* government revenues than with attracting petroleum companies to undertake exploration. For example, some governments have sought to extract large signature bonuses for exploration contracts in high-risk areas or have imposed rigid, unrealistic expenditure requirements. Model contracts employed by governments have frequently been formulated with a view to satisfying domestic ideological objections to foreign investment in petroleum production, or to protecting the monopoly status of politically powerful state petroleum enterprises. Both these

motivations apply to Argentina, Brazil, and Peru, among others. Although there has been some improvement in contract terms, e.g., Brazil, with a consequent rise in exploration activity, many years of potential production have been lost, at enormous cost to the country.

The analysis of contract terms employed by the nine countries in the case studies discussed in chapters 5–10 plus references to other countries with similar contractual arrangements showed that most fiscal systems are *regressive*, or at best *proportional* rather than *progressive*, as measured by the relationship between the government's share of net revenues and the quality of the field. A progressive fiscal system is more likely to provide an acceptable IRR on high-cost/low-volume fields than a regressive system, and therefore not only encourages the development of such fields when discovered but also raises the risk-corrected IRR of a prospective exploration investment. Regressivity of a fiscal system is produced by large signature bonuses, high royalties, and production-sharing arrangements; while progressivity is produced by a heavy reliance on net profits taxes. Not only will a better structuring of fiscal terms in petroleum contracts increase the attractiveness of projects to prospective investors, it will also tend to increase the expected net present value of government revenues as well.

Most contracts provide for lower output or revenue shares for the contractor from higher incremental production from areas under contract. Such arrangements not only reduce the expected NPV from an exploration investment, but also discourage the development of additional production from higher cost fields when they are discovered. Graduated royalties have a similar effect on the efficiency of production.

Do Existing Agreements Give Too Large a Share of Economic Rents to the Government?

This question must be answered first of all in terms of the objectives of the host government and the stage of exploration and development of a country's petroleum resources. In the case of a country that has been fairly well explored, the government may choose a lower rate of development of its petroleum resources and a higher share of net revenues. But most OIDCs, whether producers or nonproducers, have not been adequately explored and their economic welfare lies in expanding their reserves rather than in obtaining a higher share of the rents from fields with known reserves. Even though a substantial amount of exploration has been done on the mainland, as in the case of Argentina, high-risk offshore exploration may constitute the best opportunity for expanding reserves.

Although there are a number of factors that affect the attractiveness of a country for exploration by foreign companies, including contract provisions that are not directly related to the sharing of net revenues, the case studies provide a number of examples of increased interest on the part of international petroleum companies following a liberalization of contract terms. It is true that the announcement of significant oil discoveries is powerful bait for attracting petroleum companies, but a combination of few or no discoveries and harsh contract terms will elicit little interest in exploration.

Not only is it possible for the government to increase its share of rents from new contracts negotiated after discoveries have been made and interest in exploration has increased, but it is possible to structure contracts to establish different terms for individual areas with differing degrees of risk based on geologic knowledge of the area. If no reserves have been proved in certain areas, contracts might be shaped with a view to stimulating maximum interest in exploration, say in high-risk/high-cost offshore areas, or in jungle areas such as the Amazon and eastern Peru.

What Should Be the Role of GOEs in Oil and Gas Exploration and Production Programs?

GOEs are almost universal in developing countries with petroleum production and their activities include exploration, production and refining, and negotiating contracts with domestic and foreign companies. In many producing countries, including Argentina, Brazil, and India, the GOEs produce most of the country's oil and are responsible for most of the exploration activity. Although the exploration and production activities of the GOEs should be continued, governments should encourage foreign investment to maximize the exploration and development of the country's petroleum reserves just as they should utilize foreign investment for economic development in all productive sectors of the economy.

Of particular importance in a national hydrocarbons program is the creation of a market and transportation for natural gas, which substitutes for petroleum in many uses and can be exported in the form of liquefied natural gas. As noted in the case studies, many petroleum contracts have been faulted because of the absence of satisfactory arrangements for natural gas. Since in many cases exploratory wells yield gas rather than oil, the possibility of marketing gas at a profitable price is an important factor in attracting petroleum companies and will affect the expected IRR from an investment. This should be an important function of the GOEs.

GOEs should supplement their own activities to the maximum extent possible by utilizing the technical and financial resources of foreign petroleum companies, and foreign companies are the most important conduit for the transfer of such resources. The external debt crisis experienced by many LDCs in 1982–83 has limited the capacity of GOEs in some countries to maintain or expand their exploration and development activities.[2] The efficiency of a GOE should be judged on the basis of its success in mobilizing external resources for the country's national petroleum and gas programs rather than simply on the basis of its own performance in exploration and development.

[2]Press reports indicate financial constraints on GOEs in Argentina and Brazil. See *Petroleum Economist*, May 1983, p. 185; and *Oil and Gas Journal*, March 7, 1983, p. 32.

U.S. Government Policies and Foreign Petroleum Investment

The single most important U.S. government policy area affecting the investment of U.S. petroleum firms in developing countries is tax policy, particularly that relating to crediting taxes paid to foreign governments against U.S. tax liabilities. U.S. tax policy has been substantially tightened since 1977 in terms of the method of calculating foreign source taxable income and the eligibility of foreign taxes for crediting against U.S. tax obligations. The alleged position of the IRS is "neutrality" in the sense of neither encouraging nor discouraging U.S. foreign investment by means of the tax system. It is a very complex process to establish the criteria for neutrality in tax policy and an evaluation of IRS regulations with respect to this objective is beyond the competence of the author, especially given the conflicting positions of tax lawyers. An issue more relevant to this study is whether U.S. tax policy should favor U.S. petroleum investment in all or selected non-OPEC LDCs, or even in certain OPEC LDCs outside the Middle East, e.g., Indonesia and Ecuador. Tax discrimination in favor of foreign investment not only involves a measure of discrimination against domestic investment, but is in effect a form of tax subsidy that can only be justified in terms of the expected social benefits. Increased petroleum production in non-OPEC countries outside the Middle East can be shown to be of potential benefit to the United States for reasons of world supply security and as a contribution to the economic development of third world countries. Nevertheless, the justification of a tax subsidy in terms of social benefit–cost analysis is exceedingly difficult. Moreover, any selective tax benefits would probably require the negotiation of bilateral tax treaties which involve a long period of time to negotiate and even longer to obtain congressional approval. There are, however, changes in IRS regulations within the framework of existing tax law that could remove obstacles to foreign tax creditability arising from the nature of foreign tax laws without compromising the principle of tax neutrality or of providing a tax subsidy. This avenue should certainly be sympathetically explored as a means of promoting U.S. petroleum investments in countries where the issue has arisen, such as Guatemala.

The other major area of U.S. government policy for promoting foreign petroleum investment has to do with OPIC investment guarantees. For reasons presented in chapter 11, the practice of insuring high-risk investments by providing compensation limited to the actual unrecovered value of the investment has great limitations as an inducement to such investment. However, there are types of petroleum investment for which OPIC insurance has been attractive and presumably has a positive influence on investment decisions. OPIC should be given greater flexibility for shaping insurance programs that will induce petroleum investment in non-OPEC LDCs.

The Role of International Assistance Institutions

The UN has recognized the need for increased exploration in the OIDCs, but the approach of the UN General Assembly has been to pass resolutions proposing the transfer of large amounts of financial and technical assistance to the governments of third world countries for supporting activities of state petroleum enterprises. These resolutions have been tied to an ideological bias against foreign investment in mineral resources by reference to several UN resolutions relating to the sovereignty of third world countries over their natural resources. This is nonsense. In virtually all third world countries, the minerals in the subsoil belong to the state, and the governments exercise full control over the exploitation of these resources.

Somehow sovereignty over natural resources has been equated with socialism, i.e., the exploitation of resources by state enterprises. It is not surprising that most industrial countries have not been willing to contribute large sums to international agencies for financing investments that could be undertaken by international petroleum companies with large financial resources, including the partially or wholly government-owned petroleum enterprises of Britain, France, and Italy! Moreover, most investments by international petroleum companies in the third world are made in the form of contracts with state petroleum enterprises that provide the latter with ownership of the petroleum and facilities for production and a substantial measure of control over operations.

As discussed in chapter 12, the World Bank group has made a number of loans for petroleum exploration and development. The World Bank group can play a useful role in promoting petroleum development in two ways: (1) by providing financial and technical assistance to poor countries for geological mapping, limited seismic exploration, evaluating petroleum prospects and formulating exploration strategies; and (2) by serving as a catalyst for attracting foreign and domestic equity and loan capital for high-risk exploration and development. The World Bank and IDA have made a number of loans in support of (1) above, which is an important step in formulating a program for negotiating contracts with petroleum companies. However, only a few of the larger loans made to state petroleum enterprises can be regarded as serving the catalytic function indicated in (2) above. In view of the limited resources of the World Bank group and of regional international financial institutions such as the IADB, it is questionable whether these institutions should be making large loans to state enterprises for their direct operations unless it can be shown that financing for petroleum exploration and development is not available from private international sources. There may be a few cases where this is true, but international petroleum companies deny that they are unwilling to make risk investments in third world countries, even where there is little prospect of producing more than enough oil to satisfy the domestic market for the foreseeable future. Under these circumstances it is understandable that in 1981 the U.S. government opposed a proposal by World Bank president Robert McNamara for the creation of a special energy fund to be administered by the bank, since most of the loans would probably go to state petroleum enterprises.

The above observation is not to suggest that

all state petroleum enterprises are inefficient since, except for being unable to provide the most advanced technology for offshore exploration and development, several of them have been able to perform successfully. The point is that the World Bank group should not provide large amounts of financing to these enterprises in circumstances where private capital is available. Moreover, direct financing of the exploration and development operations of state petroleum enterprises appears to be at variance with the decision of the World Bank's executive directors of July 1977 to expand the lending of the World Bank group for fuel and nonfuel mineral resources in order for it to function as an "active catalyst" in promoting the flow of private international capital.

The catalytic function can be exercised by loan and equity investments which constitute only a small portion of the total investment, but which serve to provide a "presence" of the World Bank group in a contract. Since the IFC is empowered to make equity investments and has considerable experience in structuring joint private–state ventures, consideration ought to be given to enabling the IFC to play a major role in promoting the negotiation of contracts for petroleum exploration and development. The presence of the IFC could be an important factor in safeguarding petroleum companies against serious contract violations or expropriations.

Implications of Recent Events

Since the case studies in this book were prepared, there have been several events that tend to emphasize the conclusion that non-OPEC LDCs should make greater efforts to attract foreign investment in their petroleum sectors. First, the world price of oil declined in 1982–83 by 15 to 20 percent below the 1981 level, and world oil supplies became more abundant. Second, expectations for a continuous rise in oil prices over the next decade or so have changed substantially so that investors are less confident that, if they make a petroleum investment today, real oil prices will be significantly higher five or ten years hence when the projects come on stream. Third, exploration budgets of major international petroleum companies have declined as a result of reduced cash flow and there has been a decrease in their exploration and development expenditures and commitments to the OIDCs. This has meant that only the more promising prospects are being considered for investment and companies are unwilling to make large commitments in the form of bonus payments and minimum exploration expenditures for high-risk ventures.

A fourth development affecting the need for foreign investment in the petroleum sector of the OIDCs is the 1982–83 debt crisis in many of the OIDCs, including Argentina, Brazil, Chile, Colombia, the Philippines, and Turkey. The bulk of the financing for exploration and development by the GOEs has come directly or indirectly from external sources, and the debt crisis has meant a sharp cutback in net foreign borrowing in many of the OIDCs. The lending capacity of the World Bank group has been strained by the extraordinary demands of its members for assistance in dealing with structural balance-of-payments problems, and a high level of loans for petroleum exploration and development cannot be expected from this source. This suggests that in the case of those OIDCs whose GOEs have been responsible for the bulk of the exploration and development activity in the past, there must be a greater reliance on foreign investment if these activities are to be maintained or expanded.

The recent decline in the world dollar price of oil has benefited the OIDCs, but this does not reduce their need to find and develop their petroleum resources. Between 1980 and 1983 the dollar appreciated in terms of other major currencies by about the same percentage as the decline in the dollar oil price, and dollar prices of exports of the OIDCs declined sharply between 1980 and mid-1982.[3] Moreover, despite the decline in dollar oil prices, the oil deficits of the OIDCs are projected to rise over the next decade.

[3]International Monetary Fund, *International Financial Statistics* (Washington, D.C., June 1982), p. 56.

Index

144

INDEX